SOME TRUST IN CHARIOTS

The Space Shuttle Challenger Experience

by

James A. (Gene) Thomas

To Scott
May God bless,
Gene Thomas

xulon
PRESS

Some Trust in Chariots
by James (Gene) Thomas

Printed in the United States of America

ISBN 1-60034-096-2

www.xulonpress.com

Some trust in chariots, and some in horses;
But we will remember the name of the LORD our God.

They have bowed down and fallen;
But we have risen and stand upright.

Psalm 20:7-8

Table of Contents

Acknowledgements

Kennedy Space Center Public Affairs
Orlando *Sentinel*
Florida *Today*
NASA Publications
Kennedy Space Center *Spaceport News*
NASA Astronaut Corps
NASA Employees
Rockwell Employees
Lockheed Employees
HOLY BIBLE, New King James Version
The Report of the Presidential Commission on the Space Shuttle Accident

I acknowledge the priceless support of my dear wife Juanita whose computer skills brought my manuscript to life. I also thank Hugh Harris, former NASA Public Affairs official, for his fine work in editing this effort.

Foreword

The Space Shuttle Challenger accident will live forever as a tragic episode in the proud history of America's space program. I had the distinct honor to be a close participant in practically all of our nation's human space programs from the final two Mercury program flights until retirement in January 1997.

This Challenger chronicle is written for three primary reasons: (1) To relate the facts from the perspective of someone closer to the occurrences than other authors who have written on the subject, and (2) To capture as best I can the events, human emotions, and feelings as a written account for my children, grandchildren, and friends, and (3) To proclaim the importance of Christ in the life of some of the people who took part in this major tragic event.

I wish to thank all of my wonderful comrades who made this adventure so amazing. The women and men of America's space programs are the greatest technical wizards in the universe. They also are a real, vibrant, close-knit family dedicated to exploring not only space near our own planet, but the furthermost reaches of our universe.

I especially thank my family for their love and support. I praise my Lord who gave me life and who created and rules the vast universe.

The main purpose of this writing is to give a personal view of the Challenger incident from a Christian's viewpoint and from someone who was very close to the events.

I dedicate this effort to my precious family: my wife, the love of my life and best friend Juanita, and my three angels, Karen, Chuck, and Wendy.

CHAPTER 1

A Memory on the Beach

On December 17, 1996 a very large section of aluminum debris washed ashore approximately one block north of the Minuteman Causeway on Cocoa Beach, Florida. Cocoa Beach is located on the Atlantic Ocean approximately 235 miles north of Miami. Around 8 a.m. George McDonough, a former government employee, called to inform me that either a large portion of the Challenger body flap or elevon had surfaced near his Cocoa Beach timeshare condominium.

"I'm sure it's a part of Challenger, Gene. I can see the tiles still there and even distinguish some of the part numbers on the metal. I think you need to get your security folks down here quickly. A crowd is gathering and there is a lot of media coverage already."

Having finished the conversation with George, I sat there still and quiet at my desk, again reflecting on January 28, 1986, the date of the historic Challenger Shuttle accident that claimed the lives of seven great Americans. After almost eleven years, a huge section of the Challenger structure had reappeared to remind the world of the tragic event! It also reminded me of my role in the event.

I quickly phoned our Installation Operations Director, former Air Force Colonel Marvin Jones, to find as I expected, that his security personnel were already closely involved in working the details with Cocoa Beach officials to recover the part and return it to NASA property. I placed a call to our Washington NASA Headquarters

Chief, Wil Trafton. Wil was shocked to learn of our new discovery after such a long period of time. I assured Wil of our coverage, involvement, and plan to get the debris back to the Kennedy Space Center. Within ten minutes, Wil called back to advise me that "the wheels" on the ninth floor of NASA Headquarters "want that debris under locked NASA control before dark" that evening. We did even better. I was informed by 1 p.m. that the two pieces of debris, now including a much smaller segment, were safe and secure on the grounds of our gated security facility.

As expected, our public affairs personnel began to call regarding requests by the media for interviews with some of Kennedy Space Center's management. Our Center Director Jay Honeycutt was on business in California, so it became my responsibility to comment to the news reporters and television teams on our reaction to this latest discovery. Before the day was over and the eleven o'clock news finished, the Challenger debris story was a lead item on every major network and front page story for most of the nation's major newspapers.

Why such an interest by the world in debris washing up on the beach? Like so many historic catastrophes, the story of the tragic explosion of Shuttle Mission STS-51L, the Challenger, will be remembered by millions of people for as long as they live.

I decided, without a lot of hesitation, that I needed to drive to the security facility and examine the Challenger debris before my first interview with an Orlando television station at 3 p.m. Public Affairs Director Hugh Harris and I drove to the site and were warmly greeted by our security personnel. They were eager to show us the hardware. On a grassy secure spot, a large blue tarpaulin covered the 14 by 6 feet structure.

As Hugh and I stepped high over the security ropes, I felt my first sense of apprehension at looking again on the remains of the Challenger vehicle. I stooped to lift the tarp slowly and when I first saw the jagged, barnacled aluminum part, blood rushed through my body. I felt a cold chilly sense of the past and was reminded that this episode of my life would always be near in my thoughts.

"Sir, it's a good idea to look at this hardware today. With all those barnacles and sea creatures attached to it, it will be stinking to high heaven in a few hours," the operations engineer in charge of handling the debris advised.

I laughed, breaking the chain of thought about the accident. What a perplexing occurrence to have a shattered remnant of Challenger suddenly appear just before the Christmas season, reviving sad memories and causing reflection on the most traumatic experience of my lifetime. I later remarked to some of my staff that, after the shock of the first impression, my engineering experience told me instantly that this was Challenger's left inboard elevon, or at least 80% of it.

Like most aircraft, the Orbiter wings each are constructed with two large movable surfaces on their trailing edges. These surfaces, or elevons, move by command of the pilot's hand controls or by computers to turn, lift, steer, and maneuver the craft while flying in the earth's atmosphere.

After so many years of close contact with space hardware, my mind is imprinted with data. These data were obviously identifying this part to me because of the size, backward slope, outward slope, and the honeycomb structural parts. Being submerged so long under the Atlantic waters, most of the structure, some tile materials, and even some imprinted serial numbers were easily identified on the elevon. It was infested with small sea creatures and encrusted with barnacles. The left inboard elevon of the Shuttle Orbiter Challenger had survived well under the sea for almost eleven years. Some scientists believe this debris could have ridden undercurrents around the entire Atlantic Ocean and been returned to the launch vicinity by effects of the Gulf Stream. To have experienced a powerful mid-air explosion, to have fallen over 50,000 feet to the choppy Atlantic waters, and to have ridden the ocean currents for so long, the aluminum and honeycomb structure looked remarkably preserved. To the experienced Shuttle engineer, it was definitely space hardware. It was Challenger hardware.

Of course, except for the curiosity of how this phenomenon had occurred, I expected no new evidence to be uncovered as to Challenger's failures. But the memories, without doubt, stirred again in the mind of each of us who were closely involved.

What events in our lifetimes do we remember most? Powerfully imprinted upon my memory is the Sunday morning in 1943 when, as a young boy, I surrendered my heart and my life to Jesus Christ. As a nine-year-old lad, I cried as my Sunday School teacher, Mr. Ed Cochran, told us about Jesus and how important it was that we all believe in Him and make Him Lord of our lives. I accepted that as the true Gospel and the course of my life was changed that very moment. I remember how I cried that Sunday afternoon! These were tears of joy and a new feeling that a higher power cared for me, a poor boy living in a time of war and fear of the future. Beyond a doubt, this is one of the most poignant memories I have. It has been a life-changing and eternity-assuring commitment to faith in God, His Son, and the Holy Spirit.

Even after long years of scientific learning, I am more convinced than ever that there is an Almighty, righteous God who rules the vast universe.

I remember distinctly the balmy May evening in 1953 when I graduated from high school with around 400 other bright seniors-Meridian, Mississippi's young men and women leaving a guarded lifestyle to face a world plagued by global cold war and a real war in the Far East. Our nation was in the final stages of the Korean War and my destiny was to go there with the 153rd Tactical Reconnaissance Squadron of the Mississippi Air National Guard to support the war. I remember feeling no certainty as to any kind of future career in that situation. Fortunately, the war ended before my unit was officially activated. My short experience as a sergeant in the Air National Guard was rewarding, but I soon realized that a military life was not my calling.

I remember so well the Friday evening in August, 1956, when I wed the girl of my dreams, the love of my life. Juanita Purvis was a classic Southern beauty, graduating with highest honors in a class of 250 students. I remember my nervousness as I stood in a white tuxedo by my beautiful bride in her long flowing gown. I remember how, as bride and groom, we feared Pastor Jack Southerland's pants would fall since he had misplaced his suspenders. He held his pants tightly with his elbows at each side. He somehow manipulated through the passages of the Bible. We later laughed about how funny our wedding would have had been if his pants had fallen!

I also remember the hot, humid day in May, 1962 when I graduated, along with many other enthusiastic graduates, from the electrical engineering school at the Mississippi State University. We were crowded into the livestock arena, nothing more than an enclosed superdome type structure, with dirt floors and no air conditioning. It was used for livestock expositions and rodeos. The rancid smell of cattle caused the hot air to reek. Nausea was the priority feeling of the day! By the grace of God, no one near my seat got sick or fainted. Undoubtedly, one of my fondest memories of undergraduate school was this smelly ceremony.

With sheepskin in hand, I ventured forth into a world where large aerospace, communications, and electronics companies were competing for electrical engineers. Along with thousands of Korean War veterans, educated with aid of the GI bill, I found that you could almost choose your job preference and where you wished to live. My offers ranged from a technical writer with Hughes Aircraft at the Los Angeles International Airport to communications engineer with Western Union in booming Atlanta, Georgia.

My 8 by 10 inch parchment simply read: "Mississippi State University hereby confers upon James Arthur Thomas the degree of Bachelor of Science, School of Engineering, together with all the honors, privileges, and obligations thereto appertaining."

For a poor boy raised in a very low medium class family, it meant success. It seems almost shameful that my going through four long years of grueling mental torture to obtain an engineering degree would not merit more plausible words or appreciative language!

Along with my birth certificate and marriage license, this sheet of paper represented the third highlight in my first 28 years of life. One of my professors once said, "Enjoy college. You will remember it for as long as you live."

There are other memories that are impressed on my mind: the births of each of my wonderful children and their subsequent marriages where I either gave away a beautiful daughter or stood up as best man with my son. It seems that you run out of kids when you finally feel comfortable with their weddings.

Despite all these cherished memories, no experience is more strongly etched upon my mind than the events of January 28, 1986, the historic day the Challenger exploded ten miles above the Atlantic coast off the Kennedy Space Center.

In the early 1990's, a major magazine conducted a national survey of the people of my generation. They were all asked the question: "What events in your lifetime do you remember most?" The top three answers were: (1) The assassination of President John F. Kennedy in Dallas in 1963, (2) The Apollo Eleven landing on the moon in 1969, and (3) The Challenger accident in 1986. It is most interesting that two of these three memorable events were associated with the U. S. Space program.

Following the events of September 11, 2001, I am certain a similar survey would find the terrorist acts against America to be among the most remembered events of people's lives. Columbia's reentry tragedy in February 2003 is certainly fresh in our hearts and memories.

This book is about the Challenger accident, written by one who was closely involved, the Launch Director. I want this to be an interpretation of events, but more than anything else, a tribute to the men, women, and machines that are the pride of the American people, pioneers who are crossing history into space ... the last frontier.

I shall remember the Challenger forever because I held the job of Launch Director for the Shuttle program, and as such, was a major participant in its history.

This book is also about the earth - the environment around the north-central east coast of Florida. I would certainly be remiss if I did not try to describe the beauty of this land and its history - some of the wonders of the men and beasts who inhabit this semiparadise.

This account of how I lived out that episode in history and my interpretation of the events leading up to that incident is taken from my memory, my personal notes, my conversations with other participants, and from many documented accounts by people of the press and other media. It would be impossible to mention every person who has contributed to America's space efforts; they all, however, deserve special recognition.

Writing of the Kennedy Space Center, its history, environment, weather, etc., is simply to set the stage for the exciting things accomplished at the installation.

To the best of my knowledge, events are described exactly as they occurred, and my interpretation of other people's actions and words should be considered for the honesty with which I present them. This is a personal accounting of what preceded and occurred during the Challenger launch countdown.

CHAPTER 2

Shake, Rattle, and Roll

The earth is the LORD's, and all its fullness, The world and those who dwell therein...

Psalm 24:1

The Kennedy Space Center is America's gateway to the universe. The center is built on a barrier island and consists of 140,000 acres of semitropical greenery interspersed with gigantic processing facilities and launch structures. Only about five feet above sea level, this 34 mile long and 10 mile wide island is bounded by the Atlantic Ocean, Indian River, and Banana River. Indian burial mounds on Merritt Island attest to the existence of Indian tribes who inhabited the island some 3,000 years ago. Today this plush environment is inundated with orange groves, palmetto, Australian and southern pine, scrub brush, a variety of oak, palm trees, and the noxious Brazilian pepper tree. KSC, a national wildlife refuge, is home to many endangered wildlife species including the bald eagle, manatee, scrub jay, peregrine falcon, brown pelican, and several varieties of sea turtles. A silver-haired Audubon Society member was heard to say, as she sighted a particular species of water fowl near KSC's Mosquito Lagoon, "They are all out here today showing their glorious colors. What a paradise for our feathered friends to enjoy."

The National Aeronautics and Space Administration chose this remote area to build the huge Apollo launch facilities from which man would depart from earth on the lunar landing missions. In May 1961, President John F. Kennedy challenged our nation to fly a man to the moon and return him safely to the earth. The program was Apollo, the vehicle was the Saturn, and a launch site was needed. Dr. Kurt Debus, the first Kennedy Center Director, was a member of a task group to find a launch site for the mighty moon rockets. Along with Merritt Island, Florida, the group considered sites in Hawaii, Texas, and California, as well as islands off South Carolina and Georgia and in the Caribbean. For many reasons, some good, some later to be detrimental to launch success, the group chose Merritt Island over the other sites. Proximity to the existing launch and range facilities around Cape Canaveral was a definite asset for Merritt Island. The ability to launch over uninhabited land and ocean was a plus for the Florida east coast. The amount of acreage was more than enough and the semi isolation was a contributor to the selection. In partnership with an outstanding contingent from the U. S. Army Corps of Engineers, NASA carved this giant launch complex out of pristine wilderness. Bridges, canals, huge buildings, roads, and utilities were built through swamp and vegetation.

Most impressive of the gigantic facilities is the Vehicle Assembly Building, one of the largest buildings by volume in the world. Here the Saturn V Apollo launch vehicle would be mated in stages to be transported to one of two launch pads. The building of this giant complex is a model for civil and mechanical engineering, a story in itself. Easily seen from the air for hundreds of miles, the structure is more than 700 feet in length, over 500 feet in width, and stands 526 feet high. Storytellers swear that on certain days when the humidity, temperature, and dew point reach a critical stage, rain forms high in the lofty rafters of the VAB and precipitation occurs. I have never witnessed this phenomenon, and until I do, I will consider it to be a tall tale. I have on numerous occasions observed the top portion of the building obscured by low-hanging clouds. I have heard hundreds of guests gasp in amazement at their first look upward from the floor of the VAB's high bay. It is that moment when the height to the roof impresses you. Most people respond in awe when they tilt their heads further and further back to see the highest point.

All of the other facilities are massive, from the crawler/transporter to the launch pads. In 1995, the American Society of Civil Engineers published its list of the Seven Wonders of the United States. This list included the Golden Gate Bridge, the Hoover Dam, the Interstate Highway System, the Panama Canal, the Trans-Alaska Pipeline, the World Trade Center, and I'm proud to say, the Kennedy Space Center.

After the successful Apollo missions to the moon and Apollo-Soyuz missions in cooperation with the Soviet Union, NASA decided to modify several of these launch facilities to process and launch the Space Shuttle. Two new major facilities built for Shuttle were the 15,000 foot Shuttle Landing Facility, a runway twice the length of an average airport runway, and the first of three Orbiter Checkout Facilities, sophisticated hangars to provide ground checkout space for the Shuttle Orbiters. These modern, high-technology facilities are a far cry from the huts and edifices of the Indian tribes who inhabited the area for thousands of years.

The area around Merritt Island, Brevard County, has served a wide range of settlers and explorers. Beginning with the European explorers, Ponce deLeon, Hernando Desoto, and Amerigo Vespucci, the area has been a route through Florida for adventurers in search of new frontiers. Cape Canaveral was actually discovered by Vespucci, from whom our America received its name. Little did President Harry Truman know in 1949, when he established a proving ground at Cape Canaveral, that man would journey to the moon and send space probes to the planets from this region of Florida. The first launch from the Cape was a modified German V-2 rocket in July 1950. The launch tower was constructed of painters' scaffolding. A crude tar paper shack was used as a blockhouse. The rocket was totally successful, flying to an altitude of ten miles. This little projectile was a "pee-wee" compared to the earthshaking giants which were to follow. Thus began the "shake, rattle, and roll" common to residents of the Space Coast of Central Brevard County. From the success of this modified V-2 rocket built by Germans, to the Apollo Saturn V built by German-Americans, the wilderness of Cape Canaveral and Merritt Island, emerged the Kennedy Space Center, America's Spaceport. The dreamers and explorers of the 15th century have given way to a new regiment of men and women who still dream of exploring new worlds beyond the boundaries of earth's gravity. The Kennedy Space Center is truly one of the man-made marvels of the world, the primary space launch facility for our nation. Some of us who are *Star Wars* fans like to speak of the Kennedy Space Center as the foremost launch site in the "universe".

Despite its uniqueness as the starting point of man's journey to the moon, the Kennedy Space Center is barely acknowledged or recognized by the state of Florida. Even though almost two million visitors from every continent in the world visit and tour the Center each year, its place in Florida's planning is secondary. I often suspect that Florida state officials look upon Kennedy as "that Federal installation over on the east coast" and believe the best approach to handling such an entity is to basically ignore it. I had the distinct privilege of speaking at an All-American event in Cape Canaveral with Lawton Chiles, the distinguished governor of Florida at that time. In passing, I invited the noted "he coon" to come visit us at the KSC and let us show him around the Center and see the Shuttle up close. I also invited him to be a special VIP at the next Shuttle launch.

He deftly replied in his best Southern drawl, sporting his famous winning smile, "Well, Son, I've already visited out thereah. I was thereah when you boys were trying to get that big rocket ready to go to the moon. I think I've seen all you've got out thereah."

My years of higher math helped me quickly figure that the sitting governor had not been to the KSC in over 30 years and wasn't about to waste his time "thereah" again. With all due respect to our state leaders, the rich color of golden citrus, the steady flow of money from tourism, and the constant influx of retirees to the Sunshine State far exceed the appeal of the measly one billion dollars or so that the Kennedy Space Center pumps into the Florida economy on an annual basis. I would surely prefer that KSC held a more prominent place in the hearts of Florida's leaders, but short of our space workers, sadly it does not.

As the century ended and we greeted a new millennium, Florida's officials appeared to be more interested in procuring new launch opportunities for the Cape area. The threat of competition from several national and worldwide launch sites may have finally awakened the state-level authorities to the significance of Florida's spaceport. Their support is certainly key to maintaining America's leadership in the launching of future missions into space.

CHAPTER 3

Where the Eagles Fly

But those who wait on the LORD shall renew their strength;
They shall mount up with wings like eagles,
They shall run and not be weary,
They shall walk and not faint.

Isaiah 40:31

There is no other place on the earth where a lush green environment is host to a highly powerful technological activity than what exists at the Kennedy Space Center when a Space Shuttle is launched. Kennedy Space Center managers speak in their presentations of the "pristine environment" at the Kennedy Space Center. In management's view, "pristine" means pure or undisturbed. When the roar, flash, and heat of the Shuttle are spent at lift-off, the adjective "pristine" hardly applies, not to the surroundings near the launch pad. We modify our definition of "pristine" on Shuttle launch days. A powerful Shuttle launch shakes the foundation of the island's environment.

Nature lovers speak of the Kennedy Space Center as a paradise. I have grown to appreciate God's earth and its sanctity so much since being a part of the space program's day-to-day activities for many years. A pleasant drive home from work at 5 p.m. allows the observation of alligators sunning on the canal banks; wild pigs grazing on the roadside; osprey, ibis, and cranes of all shapes and sizes; a friendly manatee rolling over the waters of the lagoons between

KSC and the Cape; sometimes above all this a raft of menacing black clouds promising winds, sharp sizzling lightning, and thundering hail. The ever-changing weather, the magnificent creatures of nature, and humans have come together in this plush semitropical environment.

Dominated by thousands of long-necked cranes, the bird population on the Kennedy Space Center is quite impressive. Thousands upon thousands of birds live on or migrate to the Center yearly. Hundreds of species delight the bird watchers who visit the wildlife refuge north of the Shuttle runway. The birds most familiar and regal are birds of prey. The Kennedy Space Center is home to a special pair of winter visitors, monogamous North American bald eagles, who have built a huge nest weighing over 600 pounds. This eagle's nest is itself a phenomenon - an engineering masterpiece of nature. What God-given instinct this majestic bird must have to have designed and built this structural marvel. How could a bird select and interweave the hundreds of sticks of wood in strategic "load-points" to actually "construct" the nest? For the eagles, it was assuredly a task as difficult as construction of the huge Vertical Assembly Building. These two major projects taxed the best knowledge – in the case of our eagle friends, instinct - and physical capabilities.

Each year the bald eagle couple returns to raise eaglets and, as some employees like to believe, bring good luck to the Kennedy team. While adult eagles are content to eat one fish per day, each eaglet devours three fish daily. The bald eagle is a large bird with often as much as an 8 foot wingspan. Much larger than a commonly appearing look-alike, the osprey, the bald eagle soars at great heights while the "fisher-bird" osprey soars low to catch fish. Around the Kennedy Space Center, ospreys are known for their tendency to build nests atop deserted dead trees or on the numerous poles erected at the center to accommodate lights, cameras, or other launch equipment.

I have observed on numerous occasions the osprey, or fish-hawk, at work skimming the waters of the Banana River for its meal of the day. His powerful wingspan measures up to six feet. He snatches an unsuspecting fish with exceptional speed with his mighty claws. As most Florida visitors often do, I have mistaken the smaller majestic osprey for the mighty eagle. One Sunday afternoon while reading by the pool, I spotted an eagle about 40 feet above in a large pine. His strong claws held a struggling fish that I would count as a certain "keeper" had I caught it in the Indian River. He handled the fish with torturous adeptness as he held it by the claws, pecking its life away with the mighty beak. After a closer look, I realized the eagle having lunch was actually another fisherman osprey devouring his catch.

The eagle prefers to build its nest in live trees and our resident eagle pair resides in a tall pine west of the Kennedy Parkway between the Industrial Area and Launch Complex 39. Here they can watch over launch processing and serve as a point of interest to the thousands of tourists who visit the Center daily. If people had eyes as keen as the eagle, they could read a newspaper 300 yards away. It is fitting that the eagle is a sign of pride and excellence in America. It is estimated that Florida has almost 3000 bald eagles, second to Alaska. We like to think that a lot of eagles live and work in and around the Kennedy Space Center. I refer, of course, to the KSC team of dedicated men and women. It is my keen observation, as a watcher of eagles, that the Kennedy Space Center employs more than 10,000 people that I consider to be eagles.

One of the fastest flying birds, the peregrine falcon, is also majestic and powerful. This stately bird can fly at speeds exceeding 200 miles per hour, near the same speed at which the Shuttle Orbiter lands. With a wingspan of 3-4 feet, this exceptionally fast predator was almost extinct due to the use of insecticides and other man-made hazards. The rare sight of the bird at the Kennedy Space Center is a delight to Audubon Society members.

In addition to the bountiful birdlife at the Kennedy Space Center, there are a lot of other "critters" not so beautiful, yet very impressive.

The manatee is regularly seen in the waters of the lagoons, canals, and rivers surrounding the Kennedy Space Center. Belonging to the order of animals *Sirenia,* the manatee, or sea cow, is a playful creature roaming the coastal waters of North America. Easily weighing a ton or more, the cows seek warmer waters and are favorites of adults and children alike as they churn the water's surface. The Kennedy Space Center protects the manatee and keeps a regular accounting of each cow within its boundaries.

Early explorers, including Columbus, believed that the manatee was the "mermaid" of ancient sailing lore. The word *Sirenia* actually is derived from the Greek word for siren or "enchanting woman." No beautiful mermaid, the sea cow is more a fat, ugly creature with a gentle face much like a walrus. This air-breathing mammal has a backbone, nurses its young, and communicates through vocal sounds generally best described as squeaks or squeals. A warm-blooded animal, the manatee seeks waters with temperatures above 65 degrees F, so in a sense it is much like the migratory birds, going to warmer waters in winter. Like so many northerners who head southward when the New England leaves turn to vivid colors, these "aquatic snowbirds" establish winter residences in Florida's warmer waters. Much like the whale, the manatee must surface for air every few minutes. However, due to its capacity to fill its large lungs almost to 100%, the manatee can often stay submerged for up to 15 minutes on a dive.

A strict vegetarian, the sea cow is a threat to no other living creature. It must, however, fight for its very existence. This gentle creature spends most of the day eating the hundred or so pounds of vegetation it must consume to live. Its enemies include predators such as sharks, alligators, and humans. Other enemies are disease, cold water, boats and human vandals. In the storm and fury of the Shuttle launches, we preserve the habitat for the manatee, our gentle friend and part-time resident of our waters. An endangered species, each year more manatees die than are born. The Kennedy Space Center is a kind host to this wondrous creature.

Not so gentle, and surely one to be easily recognized, is the alligator. There are over 5000 of these wild reptiles roaming the Kennedy Space Center and the Merritt Island Wildlife Refuge, normally around the water's edge. Interesting to observe, the alligators at Kennedy Space Center can grow to be 20 feet long and weigh 900 pounds. Although big and lethargic appearing, these giants are often able to outrun a horse for 40 feet. They are especially aggressive during spring mating season and can be extremely dangerous to people. These reptiles are a definite fascination to the hundreds of Russian Space Station scientists who have visited the Center. On numerous occasions, I have observed three or four of our Russian visitors kneeling expectantly near the edge of the lake across the street from my office at the KSC Headquarters Building. Despite appropriate warning signs posted nearby, these innovative scientists often slap the water's surface with the open hand to create a loud noise in an attempt to arouse and lure a sunning alligator to come closer. I often considered having our security force warn them against this activity, but always refrained. It would be newsworthy, indeed, if an international incident occurred because a scaly Florida resident removed the arm or hand of a visiting Muscovite. The employees of KSC have grown to appreciate the danger of provoking these rustic reptiles and keep their distance.

Sea turtles, the threatened loggerhead, and the endangered green sea turtle, have nested at the Kennedy Space Center in increasing numbers over the past few years. KSC's 43 mile seashore hosts more than 6000 sea turtle nests in a given nesting period from May to September. KSC works hard to control predators that would destroy these nests, mostly wild hogs and raccoons, and egg loss is very low. Not only do we launch missiles into space, but we also yearly launch new generations of sea turtles.

Not so protected and rapidly becoming a nuisance and menace to traffic are the more than 10,000 wild hogs who roam the Center's woods and roadsides. Probably brought to the Tampa Bay area of Florida by Desoto in the 1500s, there are now estimated to be over one million wild hogs across the state. In search of select roots, eggs, and insects, these huge wild hogs are literally destroying the landscape across the Center. A drive-through of the Center's hundreds of miles of road leaves the vivid impression that the entire area is being constantly "bombed" with fragmentary explosives. Land preservation agents say these hogs will eat everything in their path. A common sight at any given time of day is a mama hog that breeds year-round and as many as twelve piglets foraging along ditches and roadways. The male hog can easily weigh up to 500 pounds. Hardly a week passes at the Center when a car, van, or truck does not suffer significant damage by hitting a wild hog who unexpectedly runs into the traffic. Not only do the hogs destroy the landscape and road vehicles, they threaten the natural vegetation and endangered species of snakes and turtle eggs. Kennedy Space Center allows hunters to trap or kill a high number of these "prolific porkers" to keep the population from overrunning the Center. Many dignitaries being entertained as a distinguished visitor to KSC have feasted on wild hog, alligator tail and swamp cabbage. It has been my pleasure to attend numerous such events. Unfortunately, I never learned to consider these dishes as culinary treats by any stretch of the imagination.

Amid the 140,000 acres of pines, scrub oaks, and cabbage palms, streaked by canals and lagoons, centered between two large rivers and part of the Atlantic Ocean, NASA has built towers of metal on giant slabs of concrete from which men and robotic probes travel to Space, to the Moon, and around the universe. The KSC is a city of sorts with food, medical care, and security inside its boundaries. I often find myself comparing the KSC to what I saw on a two week trip to tour the Russian space facilities in 1992. Were NASA obliged to copy the Communist way of life for their space workers, KSC would also contain schools, hospitals, housing, theaters, shops and all the other necessities for daily living.

At KSC there are several dozen American bald eagles in residence at certain times of the year. Added to this rare mixture of nature and technology is a work force of dedicated men and women. I often think of these men and women, full of strength and ability, as eagles. Into this area come other eagles, the crews who fly the Shuttle. The eagle is noted for its sharp vision and its powerful wings. The eagle is not only the national symbol of pride for our country, but was the emblem of the Roman Empire in its days of power. Of all the beasts that roam the lands and waters around the Kennedy Space Center, we recognize the eagle as that which best represents our people. I have always sensed not only a pride in the hearts of KSC people, but also a genuine concern and love for each other. We all shared the joy of launches, but we all suffered as one family when we lost the Challenger. We held to each other, prayed for each other, and pledged our determination to come back better and stronger.

The leadership on the space frontier will forever belong to America as long as our men and women continue to sustain this attitude of pride … eagles!

CHAPTER 4

The Kennedy Weather

Those who go down to the sea in ships,
Who do business on great waters,
They see the works of the LORD,
And His wonders in the deep,
For He commands and raises the stormy wind,
Which lifts up the waves of the sea.

Psalm 107:23-25

My favorite preacher, son Chuck, once described the drive from the NASA side of Kennedy Space Center to the Air Force side of Cape Canaveral as "totally awesome." A tremendous expanse of water on both sides of the causeway encompasses the most north-eastern part of the Banana River lagoons. Through sunglasses, the water is azure blue punctuated by scrub patches wherever the wild Brazilian pepper trees do not obstruct the view. Here creatures are free to enjoy their nature. White ibis on feathery wings hunting a meal; sea cows gracefully cutting the water's surface as if part of an aquatic show; a long-legged crane sedately strutting through the shallow waters staring at the ripples to find an unsuspecting fish; the half dozen ospreys nesting atop the public address speaker poles erected by Public Affairs to provide information for the thousands of visitors who once gathered on the causeways for Shuttle launches.

In the midst of this beautiful flat expanse of earth, sky, and water stand the gigantic concrete facilities that house and launch the Shuttle and the other unmanned launch vehicles. To the north you see Shuttle Pad B in the far distance. Pad A is a mile closer. Both are barely discernible to the unfamiliar eye. Farther south, the twin Titan pads 41 and 40 are distinguished by the tall lightning rods at the four corners of the structure. This special grid was a "second thought" design when severe afternoon lightning storms played havoc with the Air Force's most powerful rocket. As one look north to south, he sees the high-tech launch complexes, the old hangars of Cape Canaveral's "glory days," the NASA navy's two solid rocket booster recovery ships and the Atlas/Delta pads. The tree lines, though not tall, are still thick enough to hide the Cape's oldest landmark, the Cape Lighthouse. Many novice rocket fans have gazed mistakenly at the Cape Lighthouse from a distance waiting out the last seconds of a Delta or an Atlas launch. In the distance directly south, on a clear day one can see the new giant tour ship, FANTASY, docked at the Port Canaveral tourist terminal. To drive across this causeway on a sultry Florida afternoon with the black, the gray, and the white puffs of clouds decorating a panoramic blue sky make a rocket scientist's dream come true. Those of us who have spent many summers here know how swirling black clouds can suddenly bring torrents of water, hail, and lightning. These storms are so fast and powerful that you really sense the "feel" of the lightning bolts before the accompanying thunder. Many of us have a reproduction of a painting called "Cape Winds" in our collection. Few paintings have captured the wild intensity of the Cape weather and the environment of a Florida day at the Cape like this canvas by Attila Hejja. A lot of people dream of living and working in a dynamic climate like this. Through the grace of God, I was one of the fortunate ones who got the chance.

For many years, the French scientific community did intense lightning research testing at KSC during the high activity summer months. The French used instrumented aircraft and launched small rockets to attract lightning. They induced lightning to strike and be conducted to ground by a trailing wire pulled skyward by the rockets. The French chose the area because of its severe and frequent summer thunderstorms. Few years go by that some resident of the Space Coast around the Kennedy Space Center is not killed or injured seriously by lightning strikes. Scientists have found that the air around a lightning bolt reaches 54,000 degrees F, six times hotter than the sun's surface. A stroke of lightning can discharge 10-100 million volts and 30,000 amperes of electricity. The weather around the Space Coast on Florida's eastern shore is simply spectacular. The Shuttle launch pads were constructed with lightning detection and suppression as critical design requirements. NASA has built a "lightning rod" protection system by placing a huge tower above the tallest point of Shuttle Pads A and B. This lightning arrestor draws and conducts lightning strikes away from the flight hardware and the ground support equipment through large conducting wires to ground. Complementing this system is a sophisticated lightning detection and monitoring system to advise engineers when lightning has actually struck the pad, along with the intensity of the strike.

Lightning is legendary in the KSC area, the so-called center of the world's lightning activity. We actually experienced a night of severe lightning while tanking the Shuttle for the STS-8 launch. An astonishing show of spectacular lightning amazed the ground crews as we continued to press on to a successful launch in the early morning hours. I recently learned that those who fear thunderstorms have "brontophobia." A lot of KSC folks must have contracted the phobia that night!

Beyond all doubt, the dreaded hurricane is the most ominous and feared of all weather threats to Florida residents. While tornado, lightning, and rain can play havoc on launch facilities, those threats usually produce localized and limited damage. Not so with hurricanes! Hurricanes can inflict everlasting destruction in a matter of six or eight hours. Hurricanes can hit an area with devastation, move along an unpredictable course, reverse direction, and play havoc with a crippled population a second time. Hurricanes can change the environment, destroy crops and tree formations, level homes and other edifices, and disrupt power and water sources for long periods of time. Hurricane Andrew, with winds of 135 miles per hour and gusts to 165 miles per hour, devastated South Florida and Louisiana in August of 1992. The property damage from Andrew was estimated to be $20 billion.

The hurricane season begins in June and lasts until late November each year in Florida. A lot of native Floridians laugh at the weather agency's reference to the most active hurricane months as the "season." KSC engineers often jokingly remark, "I'm sure glad to see the hurricane season come to a close. We won't have to worry about those babies until June of next year!" We all know in reality that a hurricane can certainly materialize later than November and earlier than a prescribed date in June.

If there is any good characteristic of a hurricane, it is the fact that we normally have sufficient warning to prepare. Using data provided by weather satellites launched from the Kennedy Space Center, weather experts at the National Hurricane Center in Miami excel at hurricane prediction, tracking, and warning. Hundreds of thousands of Florida residents and visitors owe their lives to the talented men and women who work tirelessly to foretell the expectations associated with one of the world's most wicked natural destroyers, the mammoth circulating wind storms we have learned to give such simple names as Alice, Albert, and Camille.

At the Kennedy Space Center, NASA treats hurricanes with due respect. The Center has a tried and proven hurricane plan, a well-protected hurricane center, and a designated ride-out crew. This crew of volunteers, though few in number, is chartered to stay on the Center through a hurricane and "protect" our country's national resources. The Center is certainly aware that any protection they might provide against a 90+ miles per hour wind is nothing more than conjecture.

Our Shuttle contingency planning and scheduling always provides a "safe haven" for the Shuttle assemblies if a hurricane is threatening. "Safe haven" connotes the ability to provide inside protection in the Vehicle Assembly Building for two Shuttle stacks in the event we are required to roll one or two Shuttles back from the pad. On several occasions, we have rolled a stacked Shuttle from the pad back to the huge assembly facility because the predicted path of a hurricane brought it over or very near Brevard County when it made landfall.

Everyone who is employed at Kennedy acknowledges the destructive power of hurricanes. They are prepared to react well in advance of experiencing a hurricane's direct impact on the installation. KSC has experienced some damage to its launch complexes, but has been spared from the direct paths of hundreds of threats. How the Kennedy Space Center would fare against the forces of a Class 5 (the most powerful) hurricane is yet to be established. A Category 4 hurricane caused extensive damage to the KSC tile facility and the Vertical Assembly Building in 2004.

Why did NASA choose the KSC area to build its Shuttle site with such a history of winds, hurricanes, and lightning? The oft-given answer is that more significant factors, e.g., launch azimuth, proximity to the equator, availability of remotely located land, and the desire to avoid flying over populated areas prevailed as drivers over weather considerations. The most reasonable answer from the NASA history book is simply that other factors, i.e., cost, location, existing capabilities, and similar criteria were more favorable to the Space Coast than the questionable weather conditions.

There have been numerous reasons for Shuttle launch delays during my long association with the program. Before the early flights, NASA suffered a lot of hardware failures. Later the Shuttle experienced schedule interferences, even a sick commander, and attacks from nesting woodpeckers. The launch team experienced five nail-biting, nerve-wracking main engine shutdowns on the pad which led to many days of rework and troubleshooting. Still the number one cause of Shuttle launch delays, and in my humble opinion that which will always be the number one launch nemesis, is the weather. In the first 75 Shuttle missions, NASA experienced 50 weather delays of some nature. Launches were scrubbed for high winds aloft, out-of-limits winds at landing sites, cloud cover affecting landing and range safety visibility, fog, rain, thunderstorms, pending thunderstorms, hurricanes, and, of course, the ever present threat of lightning in the hot humid summer afternoons in Florida. It would surely have been prophetic if someone had been able to predict the Challenger accident and predict that weather would have been a significant part of it. I'm certain no one would have ever imagined a scenario where cold, chilling cold, weather would contribute to this fateful incident in history. To those of us who have lived most of our adult lives on the Space Coast, we love this land. We love its challenges and its people, and we actually love the weather. Yet, in retrospect, had weather been the deciding, driving reason to choose a Shuttle launch and landing location, the present Merritt Island area would probably never have been selected.

The great spectrum of weather conditions at the Cape range from spectacular golden-auburn sunrises to ominous dark and heinous gray, boiling afternoon cloudiness. A storm-ridden dark foreboding day on Florida's east coast can bring to mind a blustery day in the mountains of the west where snow and freezing cold are the results. Again, as it was on the day of the Challenger tragedy, the sky can be a perfect crystal blue and the temperature as bleakly cold as can be imagined.

January 28, 1986 was by a large margin the clearest and certainly one of the coldest days of my forty-three years in the Sunshine State. As some January months I have spent in Florida have been, January 1986 was a record-setter for adverse winter weather. From December to February of each year, the earth's inclination to the sun puts the sun the furthermost south. This is known as the Winter Solstice. In June, the inclination causes the sun to be furthermost north, and this is the Summer Solstice. During the Summer Solstice, every place north of the Arctic Circle will have 24 hours of sunlight and all places north of the Equator will have 12 hours of sunlight. The Winter Solstice is what causes the long, long nights in the northern parts of America in winter and the Arctic blasts of cold air to bring sub-zero temperatures to the northern states. The Winter Solstice partially accounts for the cold days in Florida in January nearly every year. We experienced one of the coldest effects of this celestial occurrence in 1986.

As the Challenger Launch Director, I was asked by hundreds of concerned space enthusiasts why we chose to launch on such a cold day. We had well defined weather restrictions as part of our launch mission rules. We violated no flight or launch guidelines. As so often is true, adverse effects upon man's endeavors lie subtly below the surface. Such was the cold weather in January 1986 and its destructive effect upon the Challenger.

CHAPTER 5

The Amazing Space Shuttle

Its rising is from one end of heaven,
And its circuit to the other end;
And there is nothing hidden from its heat.

Psalm 19:6

Trying to describe the Space Shuttle assembly is about as easy as describing the majestic splendor of Yosemite National Park or the awesome beauty of the Sistine Chapel. No words can be found to allow someone to even imagine how this massive technological man-made giant appears under bright xenon lighting without having viewed it with their own eyes. It is nothing short of spectacular and a tribute to our nation's research and development genius! Americans have designed a vehicle so powerful that it can lift 4.5 million pounds of man and machine through the rigors of the earth's atmosphere into outer space. Having experienced on numerous occasions the sight and feel of this enormous vehicle and its ground structure, it becomes almost commonplace. I have delivered hundreds of speeches to groups of people from all walks of life. Most of these people do not have technical backgrounds. They include youth groups, businessmen, church laymen and women, church congregations, and civic organizations. As I spoke more and more, I realized I needed to speak to these folks in terms understandable by the average person. After a lot of research and consultation with various Kennedy organizations, I generated some comparable Shuttle statistics that are, hopefully, easy to understand.

To envision the Solid Rocket Boosters, I ask the audiences to imagine two large grain silos, 149 feet high, 12 feet in diameter, much like the thousands of structures seen in midwestern states such as Iowa and Nebraska. The Solid Rocket Boosters are two feet shorter than the Statue of Liberty, but each weighs three times more than she does. They each consume 11,000 pounds of solid propellant per second when they are ignited at T-0. This is two million times the amount of energy burned by an average car. A single solid rocket motor produces 15.4 million horsepower, or roughly as much as 64,000 Corvettes, once the favorite auto of America's first astronauts. Two of these solid motors generate more power than thirty-two 747 jumbo jets with their four engines at takeoff power. If we converted the energy generated by the two solid rocket motors to electricity, we could supply power to 87,000 homes for a full day.

The solid motor's flame exits the rocket's nozzle at 2300 miles per hour and at a temperature of 6000 degrees F. This temperature is so hot it would not only melt steel, but boil it. The motor case insulation protects the steel case so well that the outside temperature of the solid motor is only about 130 degrees F.

Next imagine a gigantic external tank about the size and girth of the Goodyear blimp. The tank, measuring 154 feet tall and 27 feet in diameter, is attached by a strong back (attach structure) to the two solid rockets. This attach point is the Shuttle's major structural load carrying member. The external tank has a volume of over 73,600 cubic feet, the equivalent volume of six 1600 square feet houses. It contains over 500,000 gallons of liquid oxygen and liquid hydrogen to fuel the Orbiter's three main engines. Hydrogen flows from the External Tank to the main engines at 48,000 gallons per minute. The engineering mind marvels that NASA has developed a turbo pump that could drain the water from a large swimming pool in less than one minute!

The Space Shuttle main engine generates 393,800 pounds of thrust at sea level and operates at a chamber pressure of 3280 PSIA when at full power level of 512,300 pounds of thrust. The Main

Engine operates at greater temperature extremes than any mechanical system in common use today. The fuel, liquid hydrogen, is -423 degrees Fahrenheit, the second coldest liquid on Earth, and when burned with liquid oxygen, the temperature in the engine's combustion chamber reaches +6000 degrees Fahrenheit, higher than the boiling point of iron.

To this combination of external tank and two solid rocket motors, we attach the Orbiter, about the size of a DC-9 commercial airplane. The Orbiter has a cockpit built somewhat like the modern airliner of the 1970's, capable of seating up to eight astronaut crew members. The Orbiter crew cabin can accommodate a crew of ten persons in an emergency situation; as stated in the Space Transportation Systems design document. I have never given much thought to stuffing ten astronauts in the cabin, but I am sure crew contingency planners have surely given it a lot of consideration. To the rear of the cabin is a 60 feet long, 15 feet in diameter cargo bay, large enough to hold a Greyhound bus. The Orbiter cargo bay carries a 65,000 pound payload to orbit. The Orbiter has a wingspan of 76 feet and two sets of landing gear, one nose and two mains, built much like the gear for the B-1 bomber.

This Shuttle "stack" is nearly 200 feet high, weighs over 4 1/2 million pounds, and develops 54 million horsepower, equal to 18,000 diesel locomotives or 372,000 mid-sized automobiles. The electrical energy equivalent for this tremendous release of power is 40,000 megawatts of power, equal to 30 Hoover Dams. Only at Hiroshima, Nagasaki, perhaps Chernobyl, Alamogordo, and other atomic test blasts, has man released so much energy. The launch team at Kennedy Space Center does this about every six weeks. This enormous energy is expelled in just nine short minutes, 80% of it in the first 2 1/2 minutes of flight.

The Shuttle Orbiter is launched like a rocket, flies in space like an orbiting satellite, re-enters the atmosphere as a reentry craft, and lands like a glider. The Orbiter accelerates from 0 feet/second sitting on the pad to an orbital speed of 25,400 feet/second in a little over eight minutes. The solid motors burn for about two minutes, but the main engines burn longer, just a little less than eight minutes. The solids separate at an altitude of about 28 miles, and fall into the

Atlantic Ocean about 140 miles from the Kennedy Space Center. The Orbiter circles the earth every 90 minutes. It then lands with pin-point accuracy at near 200 miles/hour.

The complexity of the Shuttle flight and ground system is impossible to describe. I am convinced that the electronics, mechanics, and fluids components in the Orbiter's aft compartment and main engine area are far more complex than the electronics and similar components of the entire Saturn V launch vehicle. This 1970's vintage technological marvel has served as America's only human access to space well into the 21st century.

After many of my speeches or presentations, I'm often asked if men and women actually understand the power of this marvelous rocket system and still want to ride atop it (or more correctly, on the side of it). There is never a shortage of potential astronauts willing and able to fly on this mighty chariot. I found it interesting to read that in the early days after the introduction of the horse-drawn chariot, only the bravest, strongest, and most able men were allowed to control these new instruments of speed and power. Today, the Shuttle still demands the very best crew members. Accordingly, NASA has overcome the gender inequality and brought outstanding women pilots and scientists into the Shuttle program.

The thirst for human knowledge will continue as long as we as a nation continue to research the last frontier of space. There have always been opponents to technological advancement. The Shuttle has received, and will always receive, its share of opposition.

In 1829, Governor Martin Van Buren wrote a letter to President Andrew Jackson to protest the advent of the early railroad system. "As you well know, Mr. President, 'railroad' carriages are pulled at the enormous speed of 15 mph by 'engines' which in addition to endangering life and limb of passengers, roar and snort their way through the countryside, setting fire to crops, scaring the livestock, and frightening women and children. The Almighty certainly never intended that people should travel at such breakneck speed."

I have often wished it was possible to invite Van Buren, who later became President of the country, to experience the sound and fury of a Shuttle launch up close as a VIP guest.

Even as recently as 1970, the Friends of the Earth organization made a similar pronouncement in opposition to the supersonic transport plane: "... breaks windows, cracks walls, stampedes cattle, and will hasten the end of the American wilderness."

At the Kennedy Space Center, we often frighten fish causing hundreds of them to jump from the surrounding lagoons and birds which wing frantically away from the blast area. The Shuttle sound pressure has reportedly cracked windows along Cocoa Beach. Yet the shuttle co-exists with the pristine environment and releases its enormous energy with responsibility and pride. NASA is always attuned to the effect of every launch and keeps a well-documented account of wildlife, fish, and nearby vegetation that is impacted.

There will continue to be those who criticize our efforts to send probes to Mars, to explore the universe, and to propel humans beyond earth's confines. America, the land we cherish, was discovered because there were adventurers who were determined to explore to find new frontiers. We must never lose the zeal to be the premier nation in the forefront of space exploration.

Some trust in chariots and some in horses,
But we will remember the name of the Lord our God.

Psalm 20:7

CHAPTER 6

"Attention in the Firing Room"

At Kennedy Space Center, one often hears the words, "Attention in the Firing Room." The important function of this large, security-controlled room as the nerve center for launching the Space Shuttle demands disciplined involvement from those who enter. Still, when a special announcement is required, the Test Director will not hesitate to get the occupants' attention.

"Attention in the Firing Room! All console operators please refrain from performing BITE status reads until the Master Console has reconfigured the launch data bus."

Filled with space jargon, the announcements mean little to the Firing Room visitor. To the console operators, it commands discipline. Probably one of the most difficult aspects of the Shuttle program to describe is the Launch Control Center Firing Room. This large complex can be considered the brains of the Shuttle Launch System. To work in a Firing Room requires a special kind of person, one willing to train for long hours. Firing Room operators take responsibility for other people's lives and a lot of expensive flight and ground hardware. There are four of these gray-walled nerve centers in the Launch Control Center: two primary launch Firing Rooms, one back-up Support Firing Room used for all launches, and a smaller Firing Room used primarily for systems testing of the Shuttle flight elements. These launch facilities were named "Firing Rooms" primarily because they are Shuttle versions of the "Firing" rooms used to launch the Apollo Saturn vehicles. They are located in precisely the same space as the old Apollo rooms. The Apollo Firing Rooms were aptly named by the Germans. The high ceilings in the rooms are approximately the size of an average high school auditorium in the average sized American city. There are fifteen to twenty systems consoles facing the huge array of two inch thick plate glass windows which afford a panoramic view of the two launch pads located five miles to the east, only 300 yards from the Atlantic Ocean. Outside these windows are large metal louvers about the size of garage doors. The Launch Director can adjust the angle of these louvers 180 degrees in order to shade the room from the sun or allow a better view of one of the launch pads. Most console operators cannot see the launch ascent until the vehicle has cleared the pad to a considerable altitude. With their backs to the windows and facing the systems consoles are three rows of test directors and element test conductors. These positions are manned by 2/3 contractor personnel and 1/3 by NASA government engineers. On the lowest level of these three rows sit the external tank and SRB element test conductors, the medical doctors/aeromedical, and safety personnel. In the middle row, the NASA Shuttle Test Directors, the Orbiter Test Conductor, the Landing/Recovery Director and the Test Support Manager are the key people who perform the significant test management functions. These people are superstars, cool and precise, and dedicated to test discipline which is paramount to a smooth launch countdown. On the top row, with the best seat in the house, is the Launch Director's position. The Launch Director console level is about six feet in elevation above the console floor level. At the moment of launch, T-0, the Launch Director can simply turn to the window and enjoy the best show in town. With him on this level sit the Center's top managers. At the time of the Challenger, the Launch Director, the KSC Center Director, the Shuttle Payloads Director, the Challenger Flow Director, and the KSC Public Affairs primary spokesperson sat on the top row. Of these positions, only the Launch Director had an active speaking role in the countdown process. He is the single person who actually decides whether to launch the Shuttle on a given day. To each side of the Launch Director's row are two "glass bubbles" where the "hummers" are seated. About the size of an average school bus, these two glass enclosed support rooms provide seating for the top level program and project managers. KSC managers comically refer to these Headquarters dignitaries and managers from other centers as "hummers" because someone has said of them, "When something significant happens or they hear someone say something unusual over the communications nets, they simply look real involved and say "Hmmm..." To be serious, these experts and Center Directors who sit as consultants to the launch team have played important roles in the success of the Shuttle program. It is certainly appropriate that they have a prominent position near the action on launch day. Around the walls of the two main Firing Rooms, FR1 and FR3, hang commemorative plaques of each Shuttle launch. These simple wooden shields, approximately eight inches wide and one foot high, hung after the completion of each mission, feature one of the crew mission patches under a clear plastic guard. Hanging from each mission plaque are two medallions, the first giving the Shuttle Launch date, the second the Orbiter landing date. All of these Shuttle plaques have two such medallions, with one exception. The Mission STS-51L, Challenger plaque does not have a medallion for the landing date. Columbia's last mission plaque will be hung with its landing date also missing.

To the special visitors we escort to one of the glass bubbles to get a close-up view of a firing room, it appears to be so unique, so exciting, almost "Buck Rogers" in design. To the insider, the data and information technology used in the Shuttle launch control consoles is mid-1970's design. The processor speeds and storage capacity are well exceeded today by personal desk-top computers in most offices. The thrill of working in this environment is still awesome during the launch preparations and countdown. But day-to-day, the systems engineers work long, grueling hours performing repetitive activities to prepare another set of flight hardware for launch.

The KSC Firing Rooms have provided excitement for the occupants on numerous occasions. The rooms are protected by a gaseous Halon fire extinguisher system. The Halon gas is advertised and certified to be breathable by humans. It also extinguishes fire. In the event of a Halon release, when there is critical hazardous testing in progress, engineers are expected to continue to man their consoles.

We have had a few exciting episodes when partial Halon release scared the stuffing out of a few busy console operators. But, to a person, no one has ever neglected to stay where they are needed to protect people and hardware.

Some of KSC's zealous security personnel on one occasion advised hard-nosed Launch Director George Page that they suspected drugs in the LCC Firing Rooms. They persuaded Mr. Page to let them bring the big drug-sniffing dogs into the Firing Room. These babies would sniff out the culprits real fast. So, two or three guards with big dogs on leash strolled into an active Firing Room and created a ruckus. It was not a good idea. The drug raid shocked a lot of dedicated console operators. It degraded the feelings toward management. It was somewhat irresponsible to carry out this task in such a disruptive way. If there had been strong reasons for searching the people there, it could have been accomplished outside the room as they reported in or out. If the suspected problem required a room sniffing, it would have been much better to perform it on an off-shift when the room was empty of personnel or lightly manned. Those who managed the early operations were guilty of exerting a lot of unnecessary pressure that we could have easily avoided. But, as some described it, "that's the space race!"

Each of the fifteen console groups in the firing room performs a key function at some point in a Shuttle's mission preparations. On launch day, each is manned by twenty or more of the best systems engineers in the entire world. The men and women who man these positions of expertise and command are keen, alert, and responsible. They hold the lives of a Shuttle crew and a multimillion dollar Shuttle, a national asset, in their hands. To even comprise a list of these space heroes, both contractor and government employees, would require quite a large journal. As I relate my Shuttle experiences, I often talk of some of these people, not to single out any person, but to relate some incident in which they were involved. I know that I never would have reached any magnitude of success in what I was able to do in my personal career without the support of thousands of dedicated Americans like those on the Shuttle launch team. I have been somewhat reluctant to accept awards and accolades for anything I have done because I realize that a lot of great performers out there in all walks of life make a few other people look good. Success comes not by few, but by many!

I find it almost comical when I consider the mix of engineering backgrounds that comprise the launch team engineering contingent. There are a lot of "animals" in the firing rooms during a Shuttle countdown; Florida Gators, Auburn War Eagles, Georgia and Mississippi State Bulldogs, LSU and Clemson Tigers, Kentucky Wildcats, Texas Longhorns, Michigan Wolverines, California Bruins, and Colorado Buffalo. Thank goodness some order exists through the efforts of Tennessee Volunteers, Virginia Cavaliers, UCF Golden Knights, Purdue Boilermakers, Nebraska Cornhuskers, Ohio State Buckeyes, and Vanderbilt Commodores. In spite of this balance of control, an occasional Ramblin' Wreck from Georgia Tech or a Miami Hurricane must be reckoned with. In a serious vein, men and women from all over America, from Puerto Rico, Mexico and Canada, from around the globe, play significant roles in the launch processing of the Shuttle fleet. It is a tribute to the American spirit that graduate engineers, who once yelled for their alma maters to defeat another's, can join minds and talents to form an All-American launch team.

The southeast corner of each prime firing room is the location of the first console closest to the security check point and the firing room doors. This console seats the communication and navigation system engineers who are responsible for assuring that the communication and navigation avionics, antennas, and associated processing hardware is operating properly prior to launch. Without the communication/navigation flight hardware working well, the air-to-ground voice communications, the S-band telemetry data and uplink commands, the rendezvous radar, the landing navigation aids, the television downlink and payload data processing capabilities could be jeopardized or completely lost. Everyone associated with the human space programs appreciates and expects that voice contact between the crew and ground controllers will be maintained, clear and intelligible. Likewise, the dissemination of data from the Shuttle and its onboard systems and payloads is always expected. The crystal clear color television broadcasts from the Orbiter have enhanced the world's appreciation for space missions. A lot of the hard work, headaches, and long hours are spent by each communication and navigation engineer and technician to assure the systems are "Go for Launch." This "toil and sweat" is a way of life for all the people who work each of the several dozens of systems on the Orbiter, Solid Rocket Boosters, Main Engines, and the External Tank.

North of the communication/navigation console is the payloads systems console where engineers responsible for most of the payloads are seated. As payloads change from mission to mission, so do the people who occupy these responsible positions. Behind the communication/navigation console on the second row of firing room consoles is one of the nerve cells of the launch team. Here sit the environmental control systems and the fuel cell systems personnel. Without the environmental control functions, the crew has no oxygen and nitrogen-constituted air to breathe. These engineers also provide the gases for fuel cell power generation required onboard the Orbiter to sustain electrical power. The environmental systems engineers also manage the crew's drinking, or potable, water. They control the water and freon cooling loops required to cool the Orbiter compartments and electronic black boxes. They are responsible for the cleansing and carbon dioxide washing of the onboard air so that the crew always has life-sustaining breathing capability. And of course, these engineers get the job of operations and maintenance work on the crew potty, a device somewhat like a normal home toilet. Needless to say, Orbiters return from space smelling extremely foul and in dire need of a bathroom deodorizer, sometimes almost to the point of needing a space-like RotoRooter job. Dumping of urine overboard the Orbiter via a urine dump valve has often caused hardship due to the urine freezing in the cold vacuum of space. It has also provided space travelers with snow shower extravaganzas! These messy systems are the responsibility of the same environmental engineers who provide clean air and water. Surely our dedicated environmental engineers do not deserve the moniker we jokingly assign them ... "potty engineer." The fuel cell engineers at this console provide expertise for activating, purging, and operating the Orbiter's gaseous energized "batteries" or fuel cells. These three fuel cell plants provide direct current at 28-32 volts of power for operating the Orbiter's critical onboard systems.

To the north of the environmental controls/fuel cell console sits the outstanding Shuttle propulsion team; the liquid oxygen (LOX) and liquid hydrogen (LH2) systems engineers and the Main Propulsion and Main Engine system engineers. These experts in flight and ground systems are responsible for loading the 500,000 gallons of liquid oxygen and liquid hydrogen into the giant External Tank. They battle cantankerous pumps, leaking valves, leaking interconnects, weather conditions, and hazardous gases to provide the main propulsion for lifting the 4 1/2 million pounds of Shuttle hardware into earth orbit. This is a proud team, long on experience, a dedicated crew of contractors and civil servants. One of the most significant differences between the LOX/LH2 tanking crew and most other Firing Room systems is the need for them to remain on-station in the firing room for upwards of 12-16 hours during each Shuttle countdown. These experts must start tanking preparations at T-9 hours and often remain past T-0 launch time for up to T+3 hours in the event of a lengthy countdown hold. Yet this crew is always alert and on top of all contingencies that arise. This long period of diligent countdown participation is even more grueling, due to the fact that it normally starts in the late evening hours before midnight and extends through a launch T-0 sometime in the morning prior to noon. This tanking team has camaraderie like none other in the Firing Room. They pride themselves in "gassing up" the vehicle to provide the very best "high octane" fuel available. Their success rate has been phenomenal despite mysterious leaks and attacks by the northern woodpeckers who tried to remove the critical thermal protection foam from the External Tank. These amorous winged Romeos thought the "bark" on this "big brown tree" would make a comfortable nest for their young if they could dig out a few pieces. After help from thousands of bird experts and ornithologists from around the world, the best deterrent to keeping the woodpeckers away turned out to be big plastic owls placed at strategic points on the launch structure around the tank.

Behind the Environmental Systems console sit the electrical engineers who are the "power and light" providers for the entire vehicle. They distribute the power generated either by ground power supplies (prior to launch) or the fuel cells (late in the countdown through the mission) to hundreds of systems onboard. Without the three main direct current busses and the alternating current produced by onboard converters, the entire vehicle is without power. There are a few mechanical functions that the flight crew can perform in an emergency, but if the fuel cells are lost or the DC busses go down, disaster and loss of vehicle and flight crew is imminent. There are a lot of "crusty old veterans" in this Electrical Power and Distribution group and they all would attest to their time around the system without my having to name them. They probably have more onboard flight hardware and ground support equipment hardware to maintain and operate than any other flight system. And they do it well! Nobody does it better!

Sharing the electrical console are the Range Safety engineers. The Range Safety Destruct System is a system of destruction that you never want to employ. Only once in the history of the Shuttle program has it been used. After the Challenger explosion, the Range Safety Officer used the ground destruct command capability to destroy the errant Solid Rocket Boosters over the Atlantic. Both the solid rockets and the External Tank were configured with destruct capability from the first Shuttle flight. Command destruct of the tank was later eliminated. Government and contractor experts maintain and test certify the onboard flight system, but the ground transmit radio carrier and destroy commands are strictly controlled by the Eastern Test Range, chartered to protect the public interests from stray rocket debris, fire, or propellants. The range military, civilian, and contractor personnel take this job very seriously. Early astronaut crews did not like the idea that they were riding on a rocket that could be "blown to smithereens" by some lieutenant or his cohort with their fingers simply flipping an ARM switch and then a FIRE switch. In reality, Range Safety Officers are highly skilled, thoroughly trained, specially selected, experienced and have all the other common attributes of responsible people who hold positions of such a critical nature.

To the right of the Electrical console are the men and women who control the Orbital Maneuvering engines and the Reaction Control Thrusters. From the title, it is obvious how critical these systems are. The Orbital Maneuvering System allows the crew to fly around space, rendezvous with other orbiting stations or satellites, and, most importantly, slow the Orbiter's speed to facilitate re-entry into the earth's atmosphere for a safe return to earth. The Reaction Control System Thrusters are used for steering the Orbiter. Precise movements are required to get the Orbiter into a state vector for star-tracking, rendezvous, docking, spatial orientation, and other similar critical maneuvers. Although they don't do any "maneuvering or thrusting" in countdown, the hypergolic fuels and the tank pressurizations must be within specified flight redlines. Another team of fluids experts, they have overcome leaks, hypergolic spills, fires and hundreds of hardware failures to keep the Shuttle flying. These engineers, along with a great group of hypergolic technicians, must handle and load hypergolic fuel and oxidizer onboard the Orbiter. With techs in protective enclosed suits with air provided by back-packs, the hypergolic systems are loaded under highly hazardous conditions. A bad hypergolic fluid or oxidizer leak and a suit that doesn't seal completely could mean death or toxic damage to a person's vital organs. These people are brave heroes who do this hazardous job again and again without fanfare and with little recognition.

Behind the Electrical console sit the Instrumentation and Hazardous Gas Detection engineers. All of the thousands of flight and ground measurements which tell the health and configuration of onboard and ground systems are the responsibility of this group. The data flows from the Shuttle data processors at extremely high data rates for hardware built in the mid-70s. Today they are very slow. If you have a data problem on your system, these folks are more than eager to help you. They jokingly contend that they own no measurements if they are performing correctly, but if the data is suspect, they own everything and have to troubleshoot to locate and fix all instrumentation anomalies. The hazardous gas detection system is actually a special form of instrumentation, one that uses hi-tech sensoring techniques to sample and analyze the Orbiter and external tank compartments and interconnects so that any hazardous leaks can be detected and corrected. Many times, these folks must advise the tanking team that they have a dangerous explosive atmosphere near the Shuttle and must revert flows to the Shuttle immediately. When hazardous leaks are detected, especially when the flight crew is onboard, a lot of nail-biting goes on and the "hummers" in the glass bubbles really perform like a choir. After more than one hundred Shuttle launches, the Instrumentation crew is far ahead of the power curve and reacts with expertise.

Interspersed at the OMS/RCS console and other Firing Room locations are systems experts in hydraulics, Auxiliary Power Units (APU) and mechanisms. A combination of auxiliary power units and a network of hydraulic lines, valves, and actuators allow the Shuttle aero surfaces to be maneuvered, the main engines to be gimbaled, the landing gear to be deployed, and other highly critical mechanical systems to be operated during flight. Operating at 3000 pounds per square inch pressure, a hydraulic systems leak can spray the hot red hydraulic fluid in a dangerous and destructive manner. After they are started at T-5 minutes prior to launch, the three small Auxiliary Power Units supply the 3000 PSI hydraulic pressure to the vehicle systems. Using hydrazine as a fuel, pulse sprayed on a hot catalytic bed, these highly efficient power units are not much bigger than the average lawn mower engine. But they are a great deal more complicated and sophisticated. As I learned details of all the systems and the dynamics of the Shuttle operations, I was personally amazed at the complexity of this technological marvel.

The mechanisms team is responsible for a lot of critical events during countdown. The giant swing-arm by which the crew accesses the Orbiter cabin and other swing-arms by which the fluids systems are serviced and vented must all operate precisely in the last ten minutes prior to launch. The Orbiter Access Arm is the only route of escape to safety for the astronauts once they are strapped into their seats for launch. Everyone associated with Shuttle launches breathes a sigh of relief when the Orbiter access arm/white room and the External Tank Oxygen vent hood slowly retract clear of the Shuttle stack on launch day.

The console in the northwest corner of the Firing Rooms could be referred to as the electronic brain control for the Shuttle in its configuration at launch time. This console is manned by engineers who configure the Inertial Measuring Units on the Orbiter, the flight control hardware and aero surfaces and the guidance and navigation equipment for critical prelaunch guidance. Sitting with them are the experts on the Orbiter's five main computers, the four redundant primary computers, and the single Back-up Flight System computer. Without this complement of sophisticated control electronics, the Orbiter crew would not be able to control its main engines, its aero surfaces, or super critical pyrotechnics. It would not be able to locate and position its orbital coordinates by tracking stars or communicating with ground computers, both pre-launch and during the mission. These computer experts are affectionately called "the ones and zeroes folks." They speak a new computer language unfamiliar to the engineer like me who graduated when analog computers were the high-tech machines of the day. These engineers are referred to as DIPS, derived from the Digital Processing System (DPS) and constitute some of the brightest young minds in the Shuttle program.

The last console in the firing room is my favorite. Several years of my life and a lot of blood, sweat, and tears have been left at the Integration Console, either in Firing Room 1 or Firing Room 2. Bob Sieck, once a Shuttle Launch Director, affectionately named the console in Firing Room 1 Hezekiah because of the age of the hardware. I think I did him one better in choosing a more appropriate name. I chose the name Methuselah for the Integration Console in Firing Room 3. My Bible research only refers to Hezekiah as "resting with his fathers" and does not mention his age at death. Methuselah was the Bible's longest living man, 969 years. In reality most of the hardware that is used to launch the Shuttle is outdated 1970's vintage. With the amazing breakthroughs and advances in Information Systems over the past twenty years, the names Hezekiah and Methuselah may not be as far-fetched as they might seem.

The ground computer that runs the launch software, called simply the Ground Launch Sequencer, resides at the Integration Console. So do special software programs to provide backups for other consoles and emergency procedures in the event of a pre-launch abort. The contractor's Test Project Engineers and the government's NASA Project Engineers are hand-picked from the best systems engineers in the organizations. They are people who have proven to be good communicators, integrators, cool under pressure, and ready to put in long hours to make the Shuttle safe and successful. These men and women understand a good deal about all the other Firing Room systems and are responsible for assuring test integration is performed. They assure that no system does anything that would be detrimental to another. They review and authorize all testing involving several disciplines. All paperwork is passed through the integration console for coordination. These project engineers are 24 hours per day, seven days per week watchdogs whenever the Orbiter or the Shuttle vehicle is powered.

The establishment of the NASA Engineering Project Office is due primarily to the organizational genius of a multitalented and jovial gentleman who likes to be called Sweet Ole Charlie. Charlie Mars is a legend at the Kennedy Space Center, a veteran of the Gemini, Apollo, and Shuttle programs. You can say Charlie has been everywhere and done it all. He is by far the best planner NASA/KSC ever employed. He can see future needs, dream of results, and make the dreams reality. A cowboy at heart, Charlie's ten gallon hats set him apart from the crowds. His brainchild, to establish a group of super-charged NASA project engineers (PEs), played a significant role in meeting the myriad of integration activities involved with flying Columbia's first mission.

Mars originated the Shuttle Project Engineering Office by selecting NASA Project Engineers. Project Engineers have always been selected from among the very best systems engineers who worked on the Space Shuttle. We jokingly told some of them what distinguishes a project engineer from other people. Our boastful assessment went something like this:

> "Some people watch things happen,
> Some people wonder what happened,
> Some people ask, "What happened?"
> Some people report what happened.
> Project Engineers make things happen!"

I can proudly say that almost to a person, every project engineer I have been honored to work with has had the formidable ability to see a need for action and to cause something to happen.

One of the most fascinating stories I like to relate involves what we laughingly called "the stamina factor of the project engineer." One of the hardest working and brainiest PEs in the office was a stoical young man named Warren Lackie. His ability to visualize the software and hardware systems needed for the future was second to none. A highly motivated adult, Warren never stopped dreaming, always a few steps ahead of the rest of us. On one of his many outings to watch a launch on the Cape Canaveral side of the space center, Warren Lackie was trampled underneath a "roach coach." The giant vans used as mobile concession stands around the space center were dubbed "roach coaches." Every day they roamed all over the gated space complex to provide the work force with chili, hot dogs, barbecue sandwiches, chips, soft drinks, and similar culinary delicacies. It was no oddity to find a giant palmetto bug hitching a ride on these big chow wagons - hence, the distinction of "roach coach."

Warren and hundreds of other space workers had claimed a nice grassy plot as close to one of the Cape's launch pads as allowed. Observers always stood a good chance of seeing a space rocket explode in those early days of launch activity. No one ever wanted a mission to fail in an ascent explosion. Likewise, when they did put on a giant fireworks display, no one wanted to miss the event.

Laid back, eyes focused on the launch pad, Lackie did not see the huge van backing toward his body. Nor did the van's driver suspect he was moving so directly toward the young engineer behind him. The medical report, Warren's recollection, and the account given by witnesses, attested to the fact that the rear wheels of the gigantic mechanical beast, a mobile store weighing several tons, rolled completely over his torso. Surprised and shocked after the wheels had rolled across his chest, Warren simply crawled from beneath the van, looked around in amazement, and appeared to be still healthy from the ordeal. Although the accident miraculously failed to cause bodily damage and later medical complications for Warren, the project engineering managers still used the story to brag about the invulnerability and strength of the "average" project engineer. We often reminded newly appointed project engineers that we expected them to be so robust that an incident such as that of Warren "destroying" a roach coach would be commonplace for them.

The Ground Launch Sequencer was designed by two men, Ron Landers and Don Weinberg, who had worked the earlier Apollo/Saturn V launch sequence software. However, the two people most responsible for refining the software and getting it through the stressful early stages of STS-1, the first Shuttle flight, were two super intelligent women, Gwin Sparks and Janine Pape. I will not mention many people who have contributed to the Shuttle missions for one prime reason; it is impossible to mention a lot of outstanding individuals without leaving out the names of hundreds more who are just as deserving. These two women were instrumental in writing and verifying thousands of lines of software code that controlled the critical Orbiter and Solid Rocket Booster systems. This software is critical not only during the last crucial minutes of countdown, but also in the event of an emergency such as Main Engine shutdown, range safety abort, or a last-second "hold" called by the 300 or more systems engineers with the authority to call a hold. The Ground Launch Sequencer also monitors a large set of measurements defined by the Launch Commit Criteria. If any of these super critical parameters reach predetermined redlines late in the countdown, the Ground Launch Sequencer will hold and, if programmed to do so, command predetermined functions to occur.

It took a lot of trial and error certifications before we were certain what types of "shutdowns," or aborts, we would program the Ground Launch Sequencer to perform. Coordinating the content and roles of this super critical launch software were two of the most challenging and difficult tasks I have ever been involved with. As leader of this effort, Shuttle Project Engineer at that time Bob Sieck was brilliant. His persistence to demand only critical parameters and events to be part of the Ground Launch Sequencer resulted in the error-free software program that has become the workhorse in launching all Shuttle missions over the years.

Each of the systems engineering consoles in the Firing Room has basically the same complement of electronic components. These include a color CRT monitor which digitally displays thousands of data parameters from not only the flight vehicle hardware, but from hundreds of ground support systems needed to launch the Shuttle. The keyboard associated with each CRT is especially configured so that the function keys can be used to execute ground commands, configure hardware, call for special data displays, and activate special software routines. Each console has two or three color television monitors which can be switched to select views from hundreds of strategically placed color TV cameras across the complexes. The large percentages of these color cameras are located on the structures above and around the two Shuttle launch pads. These two hazardous launch areas are more critical. Eighty percent of the hazardous operations occur at these locations. These television cameras have played a leading role in assisting incident investigators in determining the cause of serious accidents. One of these television camera recordings provided crucial data in identifying the failure of the right hand SRB joint seal just after Challenger's lift-off. Firing room engineers use color TV to find fluid and gaseous leaks, inspect parts of the Shuttle vehicle for tile damage or ice formation, and to track the activity of red crews, ice inspection team members, flight crew members, and the Orbiter closeout crew after the External Tank has been loaded for launch.

Another important capability provided each console engineer is an Operational Intercommunications System (OIS) unit to allow voice communications between personnel at all the facilities, consoles, even remote sites around the world. This highly reliable digital voice system is invaluable in critical test situations. Its crystal clear voice reception allows test conductors and engineers to conduct the operations in a disciplined, controlled manner, and provide tape recordings for replay, if necessary. The only bad feature of the system is the seemingly impossible ability to find a comfortable lightweight headset for the users. After sixteen hours of having one of the earlier versions, "the C-clamp," on your ears you are either ready for the emergency room at the ear clinic or a padded cell. Only in the last five years or so have operational headsets been available to make the users more comfortable. Even so, the new models leave a lot to be desired.

The men and women who comprise the Shuttle launch team are like most normal people you meet every day; at least most of them are. They don't look or talk like the crews who man the Starship Enterprise. Most of them don't. There is one government quality assurance inspector who claims to have been abducted by aliens and to have traveled the far reaches of the universe. He also, after his stressful journey around the galaxy, communicates with UFOs and their crew members on a regular basis.

However, to be perfectly candid, you could actually say that the people of NASA and KSC speak a different language indeed. It is the language of acronyms. Every program, experiment, payload, facility, large piece of hardware, test, organization, or installation seems to have an acronym. One of my favorite acronyms was SPARTAN, Shuttle Pointed Autonomous Research Tool for Astronomy, which flew on the Challenger mission. It sometimes seems that NASA is afflicted with a mania, acronym-mania.

I recall in the very early days of the Shuttle program how a young co-op student from Puerto Rico, Mario Gurrero, reacted to the space language we used at the Kennedy Space Center. I was chosen to mentor the intelligent young man who was between his junior and senior years of engineering school. We were active in recruiting young engineers and especially tried to keep our co-op students interested so they would come to work for NASA. I took Mario along to attend the Orbiter scheduling meetings. One day the person in charge began the daily operations scheduling meeting by saying, "We want to roll Columbia to the VAB this weekend, so the transfer aisle has to be ready. The VITT has asked us to let a C-Square go into the bay and look at the CCTV and Ku-band DA. He'll have to go through Door 44 so we will need the 534s up. They also want to do Spacelab sharp edge inspection while they are there. We'll need a mechanical tech, a QC, and an SCO."

I comprehended all he had said and made a short note to be sure one of our off shift project engineers knew about the special inspections. But when I glanced at Mario, I saw a facial expression implying total confusion and astonishment at all the magical words and acronyms that had just spewed forth. I'm sure at that moment a similar language was being spoken in a dozen or more work areas around the KSC. To the natives, this NASA language is learned as you gain experience. It seems to me that in the experiences of the space program, you pick up the jargon and use of the acronyms real easily. I don't recall having a bit of trouble learning how to speak the language. I had a lot more difficulty learning the paper systems and the protocol. But poor Mario had a hard time with NASA-ese! When the meeting was over and we were trudging along to the next one, he said, "Gene, you engineers speak an entirely different language out here. My native language is Spanish. I grew up in a family that speaks Spanish. I learned English as a second language and I speak it well. I'm proud to be a bilingual person. But that man spoke for five minutes and all I was able to understand is that a lot of work needs to be done somewhere this weekend. I'm afraid I would never learn to speak and understand the way you guys speak." I tried to reassure him that he would learn and comprehend the language in time. I still remember his look of confusion and I'm sure he held a real concern about the language. Mario never came back to work at the Center after his graduation from college. I'm sure he was offered a job in the space program at KSC as we hired all of our co-ops who wanted to work for us in those early years. Perhaps some nice starting salary from a large research company lured him away. On the other hand, who knows, maybe Mario Gurrero never really wanted to learn a third language and said "To heck with it!" I know that all government agencies use similar means of communication. But in most agencies they center on similarities in personnel, procurement, and accounting jargon. NASA stands alone in building more acronyms, using more shortened terms, and inventing more words than any government agency. It is a NASA trademark that is recognized around the world. The crossword puzzle creators now use, "A-OK," a great deal. This is a sure sign of NASA's language making its imprint on the English language. As crazy as the NASA jargon is, in the early days we also used a lot of phrases that have come to be part of America's language. We held hundreds of "dog and pony" shows, employed quite a large number of "rocket scientists" and were always interested in the "bottom line."

The men and women who compose the Shuttle launch team and those who in some way support the team are a diverse and interesting study in Americana. There are people who love to skydive regularly, a former weapons officer on a modern nuclear submarine, race car drivers, avid sailors, former jet pilots who flew combat missions in Vietnam, former multi-engine aircraft pilots who now fly reserve and National Guard aircraft on weekends, golfers with zero handicaps, reformed alcoholics, ex-professional baseball players, football stars who made All Conference at their college or university, musicians, singers, preachers, marathon runners, pistol experts, and even a couple of stand-up comedians. They come from places like Bethpage, New York; Great Falls, Montana; Modesto, California; Austin, Texas; and Adel, Georgia. They represent the eras of post-Korean War veterans, baby boomers, and computer nourished whiz kids. There are a few silver-haired old folks left around, amicably called "dinosaurs." They have the experience that can only be gained with time. They have the corporate knowledge that is difficult to come by and more difficult to retain. They have the memories of great victories: Al Shepard's and John Glenn's first flights, Apollo 11's landing at Tranquility Base, and the first Shuttle mission STS-1. They also have the memories of sad defeats: the Apollo 1 fire, the loss of great astronauts like Elliott See, Ted Basset, Gus Grissom, Ed White, Roger Chaffee, Sonny Carter, and, of course, the Challenger and Columbia crews.

The new cadre of men and women who are taking the leadership roles in the Shuttle program are young, energetic, and confident. Most of these young engineers, technicians, and quality control specialists were either toddlers or not born when Neil Armstrong and Buzz Aldrin became the first humans to set foot on another heavenly body. They use the computer like most of us use a dinner fork. The slide rule is considered a collectible to these engineers. They have new ideas and want to see new technology introduced into the Shuttle systems. They will bring the Launch Processing System and the antiquated firing room systems into the high speed data world of the 21st century. These are the people who proudly display a chromium car tag holder that simply says "Shuttle Launch Team - Doing What Others Dream" on their flashy Preludes, Mustangs, or BMWs. Of course, some proudly display a simple bumper sticker "Launch Work is Teamwork" on their pickup truck or minivan. They will keep the Shuttle flying, maybe to the the year 2020 and beyond. These dedicated space workers can be seen all over America sporting a Shuttle launch team jacket, wearing a Shuttle mission T-shirt, a mission commemorative patch, or a NASA meatball.

Early on in NASA's history, the obvious absence of women was evident with the Mercury launch crews. There was only one prominent woman close to the action, an Air Force nurse on assignment to NASA who kept a close watch on the seven astronauts' medical condition. Eventually a few women engineers entered the NASA work force during the Gemini and Apollo programs. In the early days of Gemini and Apollo testing, it was a novel and rare experience to hear a woman's voice over the intercommunication net when a test was being conducted. It was not unusual to be manning a control room console position and have someone tap you on the shoulder and say "Switch over to Channel 142. Judy Sullivan is running a test." The word would soon get out and most operators not involved in their own test procedures would switch over to listen in wonder at an actual woman's voice giving directions and "participating right alongside men." What a simpleton attitude, but one representative of the times.

Today in the Shuttle programs, women play significant roles. Women hold key positions on the launch team: test conductors, test directors, flow directors, project engineers, lead systems engineers, Ground Launch Sequencer experts. Where it was once unusual to hear a woman's voice on the intercom, today if I were to enter a Firing Room, don my headset, and not immediately hear a woman's voice, I would certainly think that something was drastically wrong. I would probably immediately call for a countdown hold until I understood who and where the women were. It is certainly true that the Shuttle launch team, as constituted today, could not successfully operate without the thousands of outstanding women who either directly or indirectly make things happen. I look forward to the day not too many years from now when I will turn on my television to watch a Shuttle launch and hear the confident voice of a woman serving as the Launch Director.

There are some people who leave you with memories that you will never forget. There is the highly intelligent, hyperactive adult test project engineer whose reputation was simply never to wear socks. No matter the occasion, church, dinner out, launch day, or whenever, Al DeLuna never wore socks. He dressed nicely and was always presentable. But Al didn't wear socks. I'm sure if he were ever to be invited to a presidential inaugural ball, he would look great in tux and tails. But he wouldn't wear socks.

Another "unforgettable moment" happened during one of the countdowns for which I was Launch Director. About 11:30 p.m. one evening when we had just completed loading liquid oxygen and hydrogen into the massive External Tank, I noticed from my lofty Launch Director's seat a handsome young engineer enter the firing room through the electronic badge check at the door. He proceeded to the Digital Processing Systems end of the Guidance and Navigation Console. So what's the big deal? Well, the black-haired man in his late 20s wore a formal pink tux, long pink tails, white dress shoes, and one of those enormous multicolored exaggerated Mexican sombreros pushed back on his head so he could see where he was walking! He stood there emotionally and demonstrably talking to the console operators, obviously relating to them all the vivid details of the wedding he had been a groomsman for about five hours earlier. He had evidently totally enjoyed the fate of one of his good friends, the groom. It was also obvious he had enjoyed the wedding reception and probably doubted his logic for leaving the party early. Suppressing a laugh and trying to present a disciplinarian's face, I got Norm Carlson's attention and said, "Get that clown out of the firing room immediately." I don't recall ever seeing this young engineer again. I often wonder if my dedicated test conductors didn't "deep-six" him.

The people of the Kennedy Space Center are indeed the key to its success. People have made the National Aeronautics and Space Administration a great government agency. NASA's people did not become great because they pinned on a NASA badge. NASA's contractors are the best private entrepreneurs because of great people. People are the key. You cannot go through the good and bad times without loving these folks. They make your bad days worthwhile. An old bumper sticker says "A bad day fishing is better than a good day working." Close anyhow! A bad day working at KSC is far better than a good day working anywhere else! NASA's top agency management comes to KSC to support launch preparations and I have heard numerous ones remark in some way or the other, "You folks are great here at KSC. You are family! You support and respect each other. No wonder you consider KSC an ideal place to work." I don't recall exactly who it was, but a wise man once told me, "If people joke with you and have a good time with you, they like you." There was at least one instance at KSC where I not only felt liked, I must have been adored because of the following prank I was victim of.

With selection to the Senior Executive Service, the senior level of federal management, you become eligible for a few paltry perks. I hesitate to refer to what the government gives its senior managers as perks. Among these is a little more money, still about one-third of what a contractor counterpart is paid at the same responsibility level. Sometimes one gets a reserved parking space which does you very little good since your job demands that you come to work early and work long hours when there are plenty of parking spaces available. You are given the opportunity to have a free executive photo made for press releases, bios, speeches, etc. One of those jokesters who the wise man told me about was a career government employee, Bruce Jansen. Bruce's forte was practical jokes, a vocation he learned from his dad who had retired after a long career as a federal government space worker. I find it hard to determine how much Bruce must have admired me since he played so many horrendous jokes at my expense. When it came time for my annual picture update, Bruce, always the nosy one learned of this photo update session from Barbara, my secretary. He misled her into letting him have one of the first of 100 or so of the new photographs when they came from the photographers.

On the way home one busy traffic-filled afternoon, I pulled up to the normal red light you always expect at the intersection of State Road 3 and SR 520, one of Brevard County's busiest traffic spots. I sat tired and relaxed, but I began to wonder why both the person in the car to my left and the person in the car to my right were looking at me strangely. I wondered if my nose was bleeding or something. I checked everything. Even the guy ahead of me was giving me a strange look through his rear view mirror. The episode was given no further thought as we got the green light. I soon forgot about it.

Almost the same thing happened a day or so later! Drivers passing by looked at me strangely. When I stopped for a red light, two or three sets of eyes were looking at my car and me. I knew they were not admiring my rusting eight-year-old Toyota station wagon! I made a mental note to be sure to look at my tires. Maybe I had a flat or low tire. Surely they would have signaled me through some sort of hand gesture.

Before pulling into my side of the garage, I got out to inspect the entire driver's side. Nothing! Then the rear end, tail-lights, rear bumper; no missing tag! Nothing! The entire right side was normal, short of the numerous scratches and fading gray paint. As I looked down at the front tag holder bracket, I saw the object of their strange looks! I realized then why they had wanted to see what I looked like! What the rubberneckers had seen in my front tag holder was my latest executive photo, a large head shot, full color, encased in a heavy clear plastic sheath to protect it and preserve its effect for all to see. I'm amazed Bruce didn't have it laminated. What a feeling of total embarrassment! I had been driving around town, albeit a small one, and the Kennedy Space Center for three or four days displaying my own photograph as prominently as possible on the old dilapidated station wagon. I can only imagine what others who saw it must have thought. Who is this guy? Is this an ego trip or not? How crude can you get? What other conclusions could they have made? Some of them might possibly have known who I was from TV and newspaper stories. They surely must have decided that I thought a lot of myself to pull such an egotistical act by displaying my picture on the front of my car.

Jokes of this nature can sure rub you wrong at the beginning, but as you cool down, you appreciate the therapy brought to a busy work environment by humor and a good laugh. The super people who comprise the Kennedy team are the kind of folks you laugh and cry with, you work and play with, you dream with - sadly, you sometimes face tragedy with.

When Shuttle launch personnel declare that they are doing what others dream about, they are certainly justified. One cool October Sunday afternoon, I toured a distinguished physician, a real space fan, on a close-up VIP visit to the Shuttle hardware and facilities. This man was the chief of trauma-related surgeons in the state of Washington. He was enthused beyond description to visit and see the marvels he had read about and admired for so long. When I showed him the Firing Room from the glass bubble viewing point, he stood fascinated as his eyes scanned the room. "Gene, I would give everything I've worked for, my medical status, my practice, and everything I have to be able to do what you folks do here in this launch control center." I can only say that his words almost brought tears to my eyes!

You surely come to admire these men and women. There is not a lot of bonding away from the job. Most everyone wants to spend time with family or other loved ones. There are other pursuits that take a lot of stress off the work force. There are little league teams to coach, golf courses to conquer, and waves to surf. These are ordinary people, not Hollywood characters. This is the greatest launch team in the world.

CHAPTER 7

The Launch Director

The Launch Director position at the Kennedy Space Center is a very responsible and respected job. Along with the Houston Flight Director, it is probably one of the two jobs most NASA engineers would select as the best position to have in the Shuttle program aside from being an astronaut. The Launch Director controls the launch process and has the final "go-no go" authority to launch. There have been less than a dozen launch directors in the 35+ years of launching American men and women into space. I was honored to have been one of the men chosen to serve in this important position.

No better job description could be written for the Shuttle Launch Director position than the words used by Astronaut Don McMonagal. Don, one of the brightest and best astronauts to have flown the Shuttle, says "the Launch Director is the one person the Shuttle commander is waiting to hear give the flight crew the assurance that the launch process has been carried out successfully. The Launch Director is the last voice the crew hears with full authority to give the 'launch ready' status to proceed."

I would be remiss if I were not to devote some words to those who have served as Launch Directors of the manned spaceflight programs. They are each in their own way unique and individualistic people. The early emphasis on qualifications for Launch Director seemed to be that he or she be a strong disciplinarian, almost a slave driver who demanded the best of the launch team members... someone who could bark commands and expect to be obeyed like a Marine Corps drill sergeant. It also seemed to be prevalent that the leader of the vehicle operations function became Launch Director by default because they were "in charge" of the operation anyway. No other leader was campaigning for this "hot seat" position. Why not let them be in charge on launch day? Nerdy engineers never wanted to be concerned with the mundane things like launch rules, blockhouse discipline, headset discipline. Let the crackerjack operations guys do what they best know how to do! Prime examples of this operations mentality led to Paul Donnelly becoming the Launch Director for the first manned missions, Project Mercury. Likewise, operations guru George Page took the reins as Launch Director for the Gemini Program. Paul Donnelly, a smiling, cigar-loving, blonde-haired Irishman, had a voice and a flair to suit the image of the first Launch Director.

Space lore has been passed along as to the unusual way that Donnelly assumed the job of Mercury Launch Director. His predecessor was known for his outbursts, his impatience with man and machine, and supposedly his love for the spirits of drink. It seems that when totally frustrated with test progress one day, the gentleman cursed loudly, threw his headset against the console, and stormed out of the blockhouse. Sitting nearby was an alert, anxious Paul Donnelly. Paul quickly seized the opportunity to pick up the headset and assume the leadership role of Launch Director. Reportedly he did so well that day that he never again relinquished the job. He went on to become the trusted and revered Launch Director for all of the manned Mercury flights.

Paul would have probably had some difficulty with a large launch structure such as Shuttle, but for a small centrally located launch crew, he was outstanding. He had a good sense of timing and a knack of being in the right place at the right time. History records Scott Carpenter's last words to John Glenn, the first American man into earth orbit, as his lift-off time neared. "Godspeed, John Glenn" could not have been scripted by George Lucas or any playwright to have been more appropriate than those three short words. What a swell of pride in God and country did this inspire! These words were the kind of sentiments Paul Donnelly often expressed as the Mercury Launch Director. Paul sat beside Carpenter that day as the Mercury Launch Director.

Paul was soon elevated to higher management roles. This opened the way for a former vehicle test director to step into the role of Launch Director for the Gemini program. A strong leader and a no nonsense engineer, George Page made things happen. He kept his own favorites, the operations engineers, very busy and those "engineering troops" on their toes. A definite type A personality, George practiced crisis management. He never let up. He never demanded less than perfection. There were no valid excuses. He drove the wagon train until the horses fell. In spite of his nature and drive, George's winning smile and dedication caused one to admire him and go the second mile to make him happy. A James Garner look-alike, George loved his role. He was in charge; he knew it; and we all knew it. When George was ruling over a blockhouse or a control room countdown, you were quiet, you were still, and you were disciplined. A lot of young engineers have suddenly had their headset voice go mute only to turn and find George Page holding the end of their communications extension cable. He had observed an engineer standing or moving around too much and had disconnected their communications to get their attention. George Page was a graduate of Penn State University, a Nittany Lion. But those who worked close to him referred to him as another ferocious animal, the Gorilla. One of George's former deputies related how he had often told several of George's subordinates that "the Gorilla wants to see you." They invariably went to George's office, and never once inquired as to whom he was referring.

George was probably the best all-around Launch Director to hold the job. He mixed "moxie," technical knowledge, and drive to the maximum. George directed all of the Gemini launches and served the Shuttle program as Launch Director of STS-1. A serious heart condition leading to open heart surgery steered George into management to serve a short term as KSC Deputy Center Director.

The Apollo program was huge. The Apollo budget was huge. The Apollo launch team structure was huge. There were probably a dozen separate and distinct launch teams comprising the Apollo stages and spacecraft elements. There was an S-I stage, an S-II stage, an Inertial Unit stage, an S-IV-B stage, an Apollo Command Module, and a Lunar Module. There were probably other stages I didn't know about. There were test directors and test support controllers. A big complex launch structure called for one ultimate voice who gave the final "go" for the huge Saturn V. This Launch Director role was far different than any of the others had been. The top Apollo man at Kennedy Space Center, Rocco Petrone, assumed the role of Apollo Launch Director. A former Army colonel, West Point graduate, former football player on the strong cadet teams, Rocco was known as the "Italian Stallion." Rocco was one of a kind. I don't think you could have raised enough cash to try to buy a smile from him. Everything was serious. He was the picture of stoicism to the outside world. Maybe to his close associates he could show a happy face. I never had social contact with Petrone, always met him under official business conditions, and I don't recall when he ever approached producing a smile. He was universally feared by his subordinates, his peers, and a lot of his superiors. It is uncertain to my way of thinking how such leadership can be effective. However, his record speaks for itself. He led the launch center through the long period of facilities building and preparation for Apollo. He led the launch team through the first Apollo missions as Launch Director. I became very disenchanted with Dr. Petrone after Challenger when I appeared before the Rogers Commission investigating the cause of the accident. Dr. Petrone, then an executive with Rockwell International, led a contingent of Rockwell employees who testified that they did not concur with the decision to launch Challenger under the extremely cold temperatures. The facts are clear on the voice tapes of the launch decision "go-no go" poll prior to launch. Rockwell made no verbal callout to hold or delay the launch. Conversely, there is a clear "go" given for launch by Rockwell personnel. The commission report substantiates these facts. I saw Dr. Petrone in a very unflattering role of a high level contractor trying to exonerate his company of liability in the aftermath of the accident. This scenario before the commission was one of the most disappointing to witness of all the various aspects of follow-on activities. The tactics that private companies will sometimes use to avoid the loss of a precious dollar often belie the integrity that they speak of in their annual reports!

Petrone, a big Spartan individual, made a distinguished top level job of Launch Director in the Apollo era. He sat at the apex of a huge pyramid of engineers, test conductors, and managers. This pyramid worked well. It had to be successful to lead such a vast, complex combination of men and machines in reaching a decision to launch three other men on a journey to the moon.

Petrone was replaced by Walter Kapryan after the first lunar landing and was sent to the Marshall Spaceflight Center as Center Director. He also served a stint as the Apollo Program Director. Walt was a short, jolly, intelligent man who ran the Apollo Project Office at the Kennedy Space Center. Well under six feet tall, I once heard "Kappy" tell Roy Lealman, an equally short-in-stature engineering chief, "Roy, I'm beginning to think you're the only person I see eye-to-eye with around here." Everyone loved Kappy and he was easy to like. He was gentle, friendly, and fun to watch in action. Unfortunately Mr. Kapryan's first launch, Apollo 12, was struck by lightning shortly after lift-off and lost all displays, telemetry, and ground contact for a brief period of time. NASA's weather constraints were not as strict and the elements were not as scrutinized for lightning potentials as they are now for the Shuttle program. After quite a scare, Pete Conrad and crew followed their recovery procedures expertly and the mission proceeded quite successfully. When the historic explosion on Apollo 13 caused the nation to hang on edge for days, it looked to Kappy as if he were jinxed. I recall the look on his face of worry and concern. For the first time, I think I began to understand the stress to which someone in a position such as the Launch Director is subjected. Kappy was not a quitter; he fulfilled the role of Launch Director well. The Apollo program ended very successfully. I'm glad Mr. Kapryan went out in style as Launch Director. I'm not sure he ever wanted to be Launch Director for Shuttle, leaving this new challenge to fall again to George Page who had served as the Spacecraft Test Director for Apollo.

George Page made the early Shuttle launches very successful by driving the launch team like the early Romans drove the early chariots. George made the launch team perform by example. He was there early in the pre-dawn hours and late in the evening. He checked things once and checked things twice! He demanded our best and we broke our backs for him. The Shuttle team rose to every challenge. Very few of the launch team members opted to leave for other work. Some of us considered changing jobs, but pride and love of the challenging work kept us at our posts. One of the ones who took the most stress was George Page. George's heart condition eventually forced him to give up the Launch Director job. This kind of result is what makes the hard work and extra pushing seem tragic in a strange way. Is a heart condition worth the success to the individual? To the likes of men like George Page, it is. In 2002 George Page died after enjoying a quite active life of retirement... one of the real heroes of NASA's space history.

George's able operations assistant, Al O'Hara, replaced him as Launch Director. Al was not one to seek the limelight and, in my opinion, was never comfortable in this "hot seat". We were good friends and fellow church members. I remember inviting Al's family to join a group of us on a Labor Day cookout at our home. Late that afternoon Al and I sat with a couple of other men around our pool and chatted. Al shared with us his uneasiness about being in such a precarious position and told me that I should think about taking the Launch Director job someday. I remarked that I didn't see myself ever wanting to serve in that capacity. I assured Al and the others that I was content to continue as the Shuttle Project Engineer, the job I really felt was where I could contribute most. Al smiled as if he knew some line of management responsibility had been established that would get me into the Launch Director position some day later. Al O'Hara served a short period as Launch Director doing a highly professional job. An experienced veteran of many of the early space projects, Al described his impression of the first Shuttle mission, STS-1: "Nothing compared to that first one. It was the most exciting, most emotional launch I've ever been involved with." Al was somewhat forced into the Launch Director's role when George Page underwent open heart surgery after the STS-3 mission. Al's period in the job marked a time of stability which set the stage for Bob Sieck, the first Flow Director for KSC, to succeed Al as Launch Director.

Bob Sieck was by far the best qualified and most suited to be the KSC Shuttle Launch Director. A smart engineer, Bob showed his mettle learning the intricacies of the Shuttle vehicles and the ground equipment used to test and launch them. He was cool under fire. He could handle an emergency situation or a management query with the same professionalism. Although he denies it, he loved the glare of the TV lights and loved to volley back and forth with the press. Bob was dubbed "Mr. Countdown" by talented aerospace writer Beth Dickey. "Launch Director blood" flows in Bob Sieck's veins.

Bob Sieck can take a car apart piece by piece and reassemble it. He worked on his own race cars and raced sportsman class across the southeastern states. He could analyze a launch situation, whether it be weather, hardware, computers, anything, and make the right decision every time. The launch team loved and respected Bob's leadership characteristics. He was the ultimate Launch Director. It was almost sad to see someone so good at a particular job be elevated to a management position. Were this a perfect world, with all due respect to the men who have done such a tremendous job, Bob Sieck would be the NASA Launch Director whenever a crew of men and women are launched into space by the United States of America!

I had the distinct privilege of following Bob Sieck as Shuttle Launch Director. I did not replace Bob; no one could have. I did bring the same line of experience to the job, having served a long time in engineering. I think it is important that men and women who become Launch Directors have experience in highly responsible positions such as Shuttle Project Engineer. I will relish the day I hear a young woman's voice giving a Shuttle crew a last few words of farewell as she serves as the first female Launch Director. It will be even more rewarding if the first woman Launch Director followed the same engineering career path as Bob Sieck and I had experienced.

After Challenger, Bob Sieck returned to assume the Launch Director's role. He was the best and the best was needed in the recovery period NASA aptly named "Return to flight." After NASA had again established Shuttle's redesign as successfully certified, Bob returned to a high-level management position.

Following Bob's second term as Launch Director, Jim Harrington, the Challenger's Flow Director, served as Shuttle Launch Director. Cool and disciplined, Jim was a natural for the job. Jim's role as Flow Director for Challenger certainly gave him the opportunity to understand the gravity of the Launch Director position. He subsequently served in an outstanding manner for a long string of successful Shuttle launches.

In early 1998, David King and Ralph Roe were chosen by KSC to become the first tandem Launch Directors for Shuttle, each to serve for a period of six missions before rotation. I know that there has been a true "passing of the baton" to the next generation because David, a good friend and comrade, is young enough to be my son. Dedicated young men like David and Ralph provided strong Launch Director leadership for a short period of time before each was chosen to fill higher level programmatic positions.

Mike Leinbach, a former Shuttle Test Director, is serving as the present Launch Director as NASA struggles with several Shuttle design deficiencies. Mike's dedicated service in this significant launch position has been impacted by the Columbia tragedy and the tank foam incident on the following Shuttle flight.

It is almost a circle of events that brought these men to this position. I shall always wonder why I held that important position when the tragic events of Challenger unfolded. I know some day in eternity, I shall ask my Maker "Why me, Lord?" No doubt the answer will be clear, gloriously clear.

CHAPTER 8

The Origin of the Orbiter Fleet

In the final years of the successful lunar landing program Apollo, President Richard Nixon appointed an elite space task group chaired by Vice President Spiro Agnew to study how NASA's future space efforts should be planned and implemented. In September 1969, the group recommended that a manned mission to Mars begin with launch from an earth orbiting space station. Access to and from this space station would be accomplished using a shuttle between the station and earth. This shuttle eventually became our nation's Space Shuttle, officially called the Space Transportation System or STS.

Rockwell (formally a major tool builder that merged with North American Aviation) was chosen to build the main element of the Shuttle, the reusable manned Orbiter. It is not commonly known that the Orbiter named Challenger was the first Orbiter built. Structural fabrication of Challenger at Rockwell International's California plant actually got underway in 1975, about a year earlier than the origin of Columbia. It became the structural test article and underwent eleven months of strenuous dynamic testing to qualify the Orbiter structural design for the rigors of launch, ascent, return to earth, and landings.

The five Orbiters that have once been part of the Shuttle Program were aptly named for famous sailing vessels. Columbia was named after a sailing frigate which was one of the first Navy ships to sail around the world. Columbia was also the name of the Apollo 11 command module which Neil Armstrong and his crew flew to lunar orbit in July 1969. This Columbia waited for Neil and Buzz Aldrin as they became the first men to walk on the moon.

Challenger was also named for a navy ship that made extensive exploration of both the Atlantic and Pacific Oceans.

Discovery was named for both the ship of explorer Henry Hudson who discovered Hudson Bay and the ship of Captain Cook who discovered the Hawaiian Islands.

Orbiter Atlantis was the same name carried by a ship operated from Cape Cod's Woods Hole Oceanographic Institute which performed ocean research over more than half a million miles of sea.

NASA decided to honor Christa McAuliffe's profession by allowing school kids to select a name for the new Orbiter built to replace the Challenger. This last Orbiter built, the Endeavour, was named by school children in Senatobia, Mississippi and was a replacement for the Challenger.

This shuttle fleet was to provide our country with a low cost access to space, going to earth orbit, and returning 100 times per year. Its early dreamers saw the Orbiter returning to earth, becoming quickly refurbished, and launching on two week intervals.

After the Orbiter Columbia was to fly four research and development flights, Challenger was to be the "first operational Shuttle spacecraft."

After the stress of Structural Test Article (STA) testing at the Lockheed/Palmdale, California plant, STA-099 was delivered a short distance to the Rockwell/Palmdale plant for rework. STA-099 was redesignated Challenger OV-099. On July 1, 1982 Challenger was transported through the streets of Palmdale and over the Antelope Valley back roads to the Edwards Air Force Base. There it was mated to a modified Boeing 747 shuttle carrier aircraft to be ferried to the Kennedy Space Center. On July 5, 1982, Challenger made its final ferry leg from Ellington AFB in Texas to KSC, its first trip to Florida.

The Challenger could easily be referred to as a space vehicle of firsts. The first Shuttle spacewalk was performed from Challenger during its first mission, STS-6, in April, 1983. The first American woman in space, Sally Ride, made her historic flight aboard Challenger in June, 1983 with Commander Bob Crippen. The first African-American astronaut to journey into space, Guy Bluford, Jr. flew on Challenger in August, 1983. Challenger was the first Orbiter to land at the Kennedy Space Center on February 11, 1984.

Another first recorded by the Challenger was the first operational flight of the Spacelab, payload bay mounted space laboratory which can be equipped to facilitate a variety of scientific and medical experiments. Monkeys first flew on a Shuttle mission aboard Challenger. The first American woman to do an EVA (Extravehicular Activity) in space, Kathy Sullivan, did her space walk from Challenger. Challenger also flew the first five person and the first eight person Shuttle crews. The first night launch and night landing of a Shuttle was performed by Challenger in August-September 1984 and commanded by Richard Truly. Noteworthy to our friends north of the border, the first Canadian in space flew aboard Challenger.

CHAPTER 9

The Astronaut Corps

Who are these who fly like a cloud,
And like doves to their roosts?

Isaiah 60:8

One of the outstanding flyers I was privileged to work with during my long career with NASA's human space programs was my good friend and hero John Young. John Young is the epitome of an aviator, an astronaut who has done it all. John was a member of the first Gemini crew, flew Apollo 16 to the moon, and walked on the lunar surface with Charlie Duke. John Young also flew the first orbital Shuttle mission STS-1 with another great friend, Bob Crippen.

We all have our favorite stories or tales to tell about people we admire. My favorite John Young story goes something like this. When the Gemini program was started, I was a young 30ish engineer still wet behind the ears, learning the ropes of space-related work as a biomedical instrumentation engineer. John and Mercury veteran, Gus Grissom, were the two chosen to fly the first manned Gemini mission into space, Gemini Three. It was my job to test, calibrate, and maintain a complement of biomedical electronics worn by the crew to monitor their physiological parameters. These included blood pressure, body temperature, respiration, and electrocardiograms. From the EKG NASA also derived a pulse rate measurement. The early Mercury crewmen hated the dreaded "biomedical crap" and reportedly feared the "docs" would discover a medical malady that would ground them or take them out of the program entirely. So for several months, the two biomedical technicians and I walked on eggshells with Gus and John to assure we were not a bother as we plastered their bodies with sensors, wires, probes, and signal conditioners. The culmination of this activity came on Gemini III launch day when it was my honor to sit in the underground concrete blockhouse at the Titan complex 19 pad and watch medical data. This physiological data was transmitted from the astronauts' bodies through a maze of telemetry networks to one of only two locations. The only people allowed to view the data were the flight surgeon in the blockhouse and the flight surgeons in the Mission Control Center. I was allowed to view the data as the technical expert who knew the hardware best and could advise the launch team of any anomaly. I was also to advise the surgeon on how the data and hardware was being displayed.

Sometime very late in the Gemini countdown, when the clock reached T-2 minutes, the power to the vehicle and spacecraft was switched from ground power to internal power. When you had progressed this far into the countdown, the astronaut crew knew it was getting down to the nitty-gritty and T-0 was not too far away. After the power transfer was confirmed over the audio net, the veteran macho Gus Grissom spoke to the rookie John Young in a cool voice. "You okay over there, rookie?" "Yeah, I'm good," replied Young. As these very words were being spoken, I noticed Grissom's "veteran heart rate" was near 120 beats per minute and John Young's "rookie" heart rate was bumping along solid around 80 beats per minute! Accordingly, the ink writing pins on the old brush recorders in the console showed Grissom's EKG pounding away like mad and Young's just easing along with a steady, cool, repetitive rhythm. I often thought, "Was John that cool or was it his naiveté from having never flown before that kept him so calm?" Anyhow, I never had the nerve to share this data with Gus, but John and I have rehashed the event more than once and laughed aloud.

Gus Grissom was a crackerjack jet jockey, one of the original seven Mercury astronauts, and proud of his coolness and bravery. The very first time I was able to get close to Gus was about two weeks before the Apollo 1 fire which claimed his life and that of Ed White and Roger Chaffee. After years of jostling me about how "evil" the biomedical systems were, Gus came into the Apollo control room one night on second shift, sat down by my side at the biomedical console with a big broad smile on his face, and laughed, "Explain to me how all this junk works!" That evening as we talked I gained a great respect for an American who later was to give his life in a tragic accident that could have easily been avoided.

Two days after the Apollo 1 fire, I had the unhappy task of entering the foul smelling, burned-out command module to document and supervise the removal of the biomedical tape recorder. We were hopeful of finding some clue on the data tapes as to the origin of the fire. I later remarked that every engineer who worked on the manned programs should be required to go inside this gutted capsule so that the significance of safety would be impressed upon them. My observing this deadly reminder of the devastating fire forever changed my understanding of test discipline. My friend Gus and two other brave astronauts, Roger Chaffee and Ed White, gave their lives to make future space machines safer. Chaffee's daughter Sheryl was to later do an outstanding job as a secretary in the Safety and Quality organization I was to direct.

John Young once asked me what I felt were the major differences between the early Mercury/Gemini/Apollo programs and the Shuttle program. After a short but sincere thought process, I acknowledged three areas I felt were significantly different. The first was that the Shuttle was at least ten times more complex and more technically challenging than the entire Saturn V vehicle and its cargo, the Apollo Command/Service Modules and the Lunar Module. Secondly, I felt the Orbiter was different in that it was designed to fly like a rocket, satellite, and airplane. We would for the first time re-fly a manned spacecraft and be brought face-to-face with our prior work and any mistakes we might have made. Thirdly, and a very significant difference, was the Shuttle astronaut corps. These Shuttle crews were a new breed of men and women from all backgrounds, some nothing short of geniuses. The best-skilled aviators in the world were in the Shuttle corps. The first mission, STS-1 saw the beginning of a practice of astronauts coming to KSC to work alongside the engineers, management, technicians, firefighters, test conductors, and quality inspectors on a full-time day-to-day basis. We affectionately called them the Cape Crusaders, the C-Squares. The association and support that we received from the Shuttle astronauts were certainly a far cry from the Mercury program where the astronauts were treated as prima donnas. Only a select few were able to even see them. With the Gemini and Apollo astronauts came the introduction of real hero-type men into the program. The crowded schedules and their tight training regimen allowed very little time to spend at the Cape working with the hands-on hardware employees. So to be able to talk to a real live astronaut and ask him or her how they felt we should conduct portions of the countdown was not only rewarding, but invaluable in adding the "pilot's point of view" to the process.

The Shuttle era was a departure from the past space programs in that we were involved daily either through direct personal contact or by teleconference with some member of the flight crew or one of the support crew astronauts. I found these people to be highly intelligent, personable, congenial, low-key, and dedicated. From the likes of the "old" veterans like John Young and Paul Weitz to newcomers like Loren Shriver, Brewster Shaw, and Dick Truly, these folks were more like your friends than your customers. Where did the new astronaut corps come from? How were they selected and who selected them?

In January 1978, 35 astronaut candidates were selected by NASA to participate in a training and evaluation program to qualify for subsequent assignment on future Space Shuttle flight crews. This group of 20 mission specialists and 15 pilots began training at the Johnson Space Center in Houston, Texas in July 1978 and completed training in August 1979, joining the active astronaut corps as Group 8. Of this group, 14 were civilians and 21 were military officers, six were women, and four were members of minorities. They joined an existing staff of Apollo and Apollo/Soyuz astronauts and seven Air Force Manned Orbital Laboratory pilots who had joined the NASA Astronaut program in August 1969. These seven MOL astronauts included my later-to-be good friends, Bob Crippen, Bo Bobko, Hank Hartsfield, and Dick Truly. Every year or so since the original Group 8 selection, NASA has called for astronaut applications and selected new groups of pilots, mission specialists, and payload specialists. Applications come from all walks of life including ranchers, farmers, school teachers, musicians, and the like. There have been less than 200 astronauts chosen in the history of the U. S. space program. The active astronaut corps usually numbers less than 100, about half of which are pilots. NASA has no weight criteria for candidates except that they meet the standard fitness requirements of military personnel. A candidate for pilot astronaut must be between five feet four inches and six feet four inches in height. Mission specialists are responsible for a multitude of scientific chores when they are in space. A mission specialist must have a Bachelor's Degree in engineering, science, or mathematics, correctable vision, good health, and be between four feet, 10 ½ inches and six feet four inches tall. The pay for astronauts is about what the average journeyman engineer or scientist earns throughout NASA. Astronauts do not receive hazardous duty pay.

Only about 3% of the applicants NASA screens for astronaut candidacy are asked to come to the Johnson Space Center for further screening. An Astronaut Selection Board comprised of astronauts, flight surgeons, and other NASA managers then reviews each invited applicant using criteria such as medical standards, interpersonal skills, teamwork abilities, and skills in problem-solving and decision-making. If you are one of the 20 or so of the 100 who normally receive this screening, you are invited to become an Ascan or Astronaut candidate. Starting pay is about $35,000 and it will rarely exceed $80,000 no matter how many advanced degrees you acquire or how many missions you command. Being an astronaut demands long hours of training and instruction beginning with a strenuous year in a candidate status. The long hours and frequent travel often impact family life. Fortunately NASA maintains a fleet of sleek T-38 jet aircraft. Primarily built as an Air Force trainer, the T-38 gives pilots an opportunity to maintain their flight proficiency. It also provides a faster, more versatile means to get to the many sites requiring astronaut participation. Those who are not pilots can hitch a ride in the other seat whenever a flight is going their way.

Pilot astronaut candidates must have a minimum of 1,000 hours time in jet aircraft in a command pilot status. These men and women must land the Orbiter on a runway like a large airplane. They will receive extensive landing experience in a modified Grumman Gulf stream II called the Shuttle Training Aircraft. All astronaut candidates receive zero gravity training in the KC-135 aircraft and many in the neutral buoyancy water tanks at the Johnson and Marshall Space Flight Centers. The training and preparation required to become a NASA astronaut is indeed strenuous and difficult. Never have I talked to a single man or woman of the astronaut corps who was not pleased to be there and wholeheartedly avowed that the hard work was indeed worth the effort to serve in this world-class elite group.

Someone recently told me jokingly that the best qualifications for applying for the astronaut corps might well be that of being a graduate of Purdue University. My latest tally of Purdue grads who have served as a NASA astronaut at one time was almost 20 including the first man to set foot on the moon's surface, Neil Armstrong.

In 1996 two KSC employees, Frank Caldeiro and Joan Higginbotham were selected as astronaut candidates. We were elated that the second and third KSC employees were so honored. Over 2800 people responded to that call for applicants and only 122 were invited to be personally interviewed at Johnson. Frank and Joan were obviously two of these. I want to paraphrase a few excerpts from the daily log which Frank so well composed from this unique experience of being tested for selection as an astronaut candidate.

"Day Three – Noon time at the psychological ward… They gave an 1100 question quiz. Questions such as, Do you hear voices? Do animals talk to you? Do you like your friends? I'm glad that I can laugh at it now that it's all over and don't feel like screaming, "Say what!"

"Day Four – the musculoskeletal exam, the neatest test of them all measures your body and skeletal dimensions and determines if you will fit into a Russian Soyuz seat. (The requirement is because the Soyuz is the emergency return from orbit for crewmen manning the Russian space station, MIR). You are then strapped to a machine that tests your muscular outputs against a resisting arm monitored and measured by computer.

"That day I also got the ophthalmology exam which was very thorough except the dilation process decreases your vision to nothing. Luckily, someone took us to dinner and read the menu to those of us who had dilated eyes.

"After a day of recording my heart activity on an electronic monitor, on day 5 we were given the neurology exam to check our reflexes. "Follow my finger, can you feel this? Rubber hammer hang here, bang there, on the knee! Slap on the face! Follow my finger! I comically named it the Curly, Larry, and Moe Test. Follow my finger… whack!"

"Day Six: I received the echocardiogram test where heart status is measured by sound and you are able to hear and observe the muscular and blood flow activities of your own heart. What a mighty, regular organ the heart is!

"Then I had the interview of a lifetime. I was asked to write an essay on why I wanted to become an astronaut. John Young asked me about the B-1 bomber that I had worked on. Bob Cabana, Chief of the astronaut corps, asked about the experimental aircraft I had built and flown.

"That evening we were cordially entertained at a traditional astronaut gathering place, a Cajun BBQ House, by Young, Cabana, and George Abbey, the JSC Center Director. What a thrill to be encouraged by these top JSC officials and to be told by Mr. Abbey we would probably fly in 1998 if we were selected.

"Last Day – We were debriefed on the physical and psychiatric test results. I guess I'm an ordinary Joe like most of the applicants. I still didn't hear any voices while in Houston! And to top off the long week, the plane I flew home on got hit by lightning about 70 miles out of Orlando!"

Frank and Joan are quite dynamic young people who, I predict, will do great things as astronauts and play a significant role in the construction of America's orbiting space station.

Not all applicants are chosen in the first attempt like Frank Caldeiro. Not all are flown on the Shuttle as pilots, mission specialists, or payload specialists. Although often given an official title, some Shuttle flyers have had the golden opportunity, for political reasons, to promote space travel as commonplace, or to enhance international relations with other foreign countries. The first foreigner to fly on the Shuttle was Ulf Merbold of Germany. He was followed by Marc Garneau, Canada, and Paul Scully-Power of Australia. We later flew a prince from Saudi Arabia, Sultan Al-Saud, and Rodolfo Neri Velo of Mexico, both in 1985. Charles (Chuck) Walker of the McDonnell-Douglas Company flew the Shuttle three times to operate an electrofluoresis experiment from which his company planned to make a commercial investment. These foreign payload specialists were skilled individuals who provided excellent payload services on their particular missions.

In 1985 Senator Jake Garn of Utah flew on Shuttle as did Congressman Bill Nelson of Florida in 1986. It was common knowledge around the Agency that NASA planned to eventually fly more civilians on the Shuttle, so the selection of Christa McAuliffe as the first teacher in space was welcomed as a natural progression. There were repeated rumors that NASA would fly a reporter, someone such as space reporter Jay Barbree of NBC, or an entertainer like John Denver, an avid space fan and supporter. One of the earliest references to flying a civilian was found as I searched through my old log books. The short note written in an entry on November 28, 1983 simply read "We came very close to flying an aerospace magazine writer on STS-5." I have no factual confirmation that STS-5 was a candidate flight for the first civilian to fly on Shuttle. I can only speculate that if NASA considered flying a newsperson, it was promotionally motivated and an attempt to show that the Shuttle was becoming an operational means of access to space. Columbia's STS-5 mission was the first four man crew; the ejection seats had been deactivated and two seats had been installed to accommodate two more astronauts.

Christa McAuliffe was chosen to be the first Teacher in Space through a long process of elimination. Ten finalists were selected by a review panel composed of University presidents, company presidents, a doctor, an actress, a former astronaut, and a professional basketball player. The ten eager teachers underwent long days of extensive mental and physical exams at Houston. The final approval for the selection of Christa was made in Washington. On July 19, 1985 Vice President George Bush announced that from the more than 11,000 who applied, Sharon Christa McAuliffe would be the teacher to go into space aboard Challenger in the coming year and that Barbara Morgan would be her back-up. I remember how thrilled my own sister-in-law Carol was to have been given a chance to be one of the 11,000 who applied. An elementary grade teacher who loves adventure, she was disappointed not to be chosen, but elated to get a nice letter from NASA thanking her for her interest in becoming the first Teacher in Space.

In the early 90s, I was asked to present a short speech and to introduce a new female astronaut at the solid rocket refurbishment facility operated by United Space Alliance inside the Kennedy Space Center. I was especially pleased to introduce the first female Shuttle pilot to be chosen as a NASA astronaut. I certainly stretched no exceptional brain power when I announced that I was "presenting the first woman selected as a Shuttle pilot who someday in the not too distant future would command her own Shuttle launch at KSC." Eileen's command flight in December 1998 proved a fitting tribute at last to the great female aviators who have never been appropriately recognized and honored.

A lot has been written about the astronauts. The versatility, diversity, and strength of this outstanding group defies description. I had the privilege to touch many of their lives in a very small way! Some I knew as friends and brothers, some only as professionals striving to do the most challenging of technical jobs. Those who became close friends will always be dear throughout our lives. To a person, I have admired each one and cherished the opportunity to have been part of doing what others can only dream about. I have often thought how I would have felt had I followed a career path that had led me to become a member of the astronaut corps. I am persuaded that the roles I played (test engineer, Launch Director, Safety Chief) were where God intended me to serve. I could go on for pages and relate anecdotes, funny stories, and significant experiences with these great people, but a simple introduction to the uniqueness of this select group of individuals was considered to be sufficient.

CHAPTER 10

The Beach House

Its official name is the Kennedy Conference Center or something along those lines. To insiders it still is and will always be the "Beach House." Most large modern organizations maintain some type of favorite out-of-the-way facility at which they hold retreats, management sessions, and special events away from the public, void of telephones, and separated from the workplace. Some organizations prefer to periodically lease a facility remotely located rather than maintain a standing obligation. Most of the major NASA centers across the country have some sort of similar facility.

For the benefit of Washington bureaucrats and budget specialists, the Kennedy Conference Center is available for training and conferencing, two activities that are NASA institutions. For the astronaut crews and Kennedy government employees, it is a refuge from the toils of space preparations. It is simply the "Beach House." Here NASA's astronauts and spouses spend a few quiet days prior to launch. Kennedy and NASA managers have entertained royalty from all over the world there, along with Hollywood and sports dignitaries, members of Congress, influential businessmen, and foreign space executives. Hundreds of NASA brainstorming sessions have been held in this old house by the ocean. Numerous major NASA decisions have been reached by NASA executives pulled away from the busy office routine to this quiet place nestled among the cabbage palms and the sea oats.

Built in the 1940s, this old rustic house elevated two stories above the sand and brush has withstood gale and hurricane force winds for decades. Modern changes have added a "downstairs" much like an above-ground basement. In a great oceanside location, the old house presents a splendid view of Shuttle Pad A to the northwest. Those who visit are serenaded constantly by the roar of the blue Atlantic surf. Above the brick fireplace, a row of bottles commemorate the souls who tossed them out to sea and those who found them at the ocean's shoreline and put them on display. I always wondered why the large majority of the bottles that wash up on the beaches from foreign shores are liquor bottles!

The significant tradition of the KSC Beach House, a memorable part of the Shuttle pre-launch experience, has always been the barbecue dinners shared by the astronaut crews and KSC government managers a few weeks prior to each launch. The astronaut crews entertained NASA's KSC managers by having the famous Texas barbecue flown in especially for the pre-launch dinners. What a tasty western treat! Hot, spicy Texas beef! Accompanied by cold slaw, potato salad, relishes, and warm buttered bread, the meal is a real diet buster. All of this is then topped off with a big piece of moist chocolate cake with chocolate fudge icing. The taste of this tempting meal is only surpassed by the fun and fellowship of sharing a good time with the men and women who fly the Shuttle and the men and women who keep it flying.

Remembered especially were instances such as the twenty minutes or so I shared talking with Shannon Lucid before she was ferried by the Shuttle to spend an extended period of time aboard Russia's MIR station. Shannon set the world's record for an American in space, 188 days, and won the hearts of millions of earthbound admirers.

These barbecue dinners close to launch became a tradition early in the Shuttle days. The dinners brought the key launch team managers close to the flight crew for a time of bonding of our minds and spirits. I only regret that every member of the Shuttle launch team could not share in the camaraderie and company of the great people who fly the Shuttle. It was always a quiet dinner with no loud jesting or revelry. Those who drink beer and wine enjoyed a drink together after a hard day at work. Those of us who stick to soft drinks enjoyed the evening just as well. I always felt a special closeness in sharing with men and women who were all conscious of the risks the shuttle was subjected to each time we launched. We rarely talked of Shuttle things; we seemed to be more interested in telling of aviators who had gone before. The early astronaut legends were always good topics at the dinner table. I could always manage to catch the attention of those at my table because I had lived through numerous experiences with astronauts of all the previous manned programs. They all seemed to enjoy hearing about the character and characteristics of those who flew into space in the pioneering programs before Shuttle. I was asked often by a Shuttle astronaut questions like: "What was Neil Armstrong like?"

There have been numerous touching and humorous instances during these Beach House barbecue sessions. As a prelude to one Shuttle mission, we met with the multitalented crew when they were at the center to participate in their countdown demonstration. A member of the crew was the oldest American to fly into space, Omega man Story Musgrave. Everyone loved Story and, because of his naiveté, he often became the object of a lot of friendly ribbing. Most of the joking appeared to simply pass right over his shaved head. This brilliant, hard-working astronaut/scientist holds seven or eight advanced college degrees, two or three at the doctorate level. Easily recognized by his bald pate, Story wears his hair in the fashion of the late Yul Brynner and Michael Jordan. Close and slick!

The evening of the astronaut barbecue, most of the KSC managers arrived 15-20 minutes prior to the crew who were practicing pad emergency egress procedures late in the afternoon. We were standing around relaxing after a long day when the crew arrived in force. Led by their commander, every member of the crew wore the traditional blue astronaut coveralls, an ID patch over the heart reading simply " STORY," their mission patch, and on their heads a sheet of latex rubber cut to look exactly like Musgrave's bald head! In an orderly fashion each crew member came through the screen door, held out their hand to shake, and charismatically spoke, "Hi, I'm Story," "Hi, I'm Story," "Hi, I'm Story," etc. Without careful scrutiny, it was almost impossible to distinguish if one of the gleeful masqueraders was the real Story. After speaking and shaking hands with all six, I realized that the brunt of their coordinated joke was missing! Though never confirmed, I remain certain that the renowned prankster Steve Hawley was the chief perpetrator of this stunt. I remember well his penchant for disguises from the STS-61C Groucho Marx impersonation. Whoever the promoter and whatever the motive, it was one of the funniest attempts at comedy I ever witnessed by a group of astronauts.

Such was the banter and atmosphere surrounding our Beach House socials with the Shuttle crews. Always light, never too technical or business related, it was a time to relax and enjoy each other's company.

Another funny episode at the Beach House occurred during the first mission in which the Russians were involved. We hosted a dozen or more Russian engineers and technicians for the STS-63 mission in February 1995 on which a significant Soviet hardware module was flown. We felt it appropriate to invite a few of their top managers to the barbecue. Totally enthused over the semitropical Florida environment, alligators, and the like, these scientists were equally captivated by the rustic Beach House a few short yards from the beautiful roaring surf of the Atlantic Ocean. The lack of bathing suits and the chilly air did not deter three or four of the Russians from wanting to strip to their typical European men's bikini shorts and hit the waves. One or two preferred "au naturel" and proceeded to prepare to skinny dip. One of our quick- thinking KSC managers deftly suggested they retain their bikini shorts due to the presence of several women at the event. They gladly agreed, but I'm sure their preference would have been to enjoy their cold ocean swim in their birthday suits!

In spite of the warmness of this traditional meal, without exception, I never left one of their early evening socials without thinking that this might well be the last time that I see this flight crew up close and alive. People might interpret such thinking as pessimism, even some lending more toward fatalism. But I did experience those thoughts on all occasions after a Beach House meal with each crew. I prefer to consider the real fear I recognized as realism. As I buckled my seat belt, a safety practice, I was always reminded to pray for the safety of my friends and America's heroes. I prayed earnestly each time I left the Beach House that my fears would never materialize, that God would protect and preserve their lives.

CHAPTER 11

Shuttle: The Early Days

If I ascend into heaven, You are there...

Psalm 139:8

The Orbiter Columbia arrived at the Kennedy Space Center in August 1978 for preparation as the first flight of a reusable manned space machine. It would probably be considered unbelievable if you told a young engineer at the Kennedy Space Center today that it took two years and eight months to process the first Orbiter. As a young fortyish project engineer, I felt as if I spent twenty of those years working 12-16 hour days, seven days a week helping prepare Columbia for launch. I had a lot of company. NASA decided to go ahead and ship the unfinished Orbiter Columbia to KSC and "finish it down there." The first Shuttle launch was behind schedule and Congress was asking why. Politics often drive decisions in the space program that sounder minds might not make. I also strongly suspected that the Rockwell work force in Palmdale, California was not overly excited to complete the orbiter and ship their livelihood out the hangar door to Florida.

So NASA shipped Columbia to the Kennedy Space Center, along with a lot of reputable managers of earlier renown, who had led other important NASA projects. The Orbiter's super critical thermal protection tiles were yet to pass all their qualification testing. We would still be required to fit thousands of tiles, bond them on the Orbiter's skin, "pull test" them, "water break" test them; you name it, we did it. I'm convinced that somewhere in some lab, some tile expert was actually "taste testing" the blasted things! What a headache and a nightmare we experienced in effecting the installation of more than 30,000 of these black and white silicon based, heat resistive tiles.

We actually were sent so much help from outside the Kennedy Space Center that Orbiter processing and tile processing often got in each other's way. A letter was finally officially released that in so many words said "Stay out of tile's way. The Columbia tile work has priority over everything else until the work is back on schedule." In spite of the priority given to tile work, a lot of dedicated engineers, technicians, and quality control personnel struggled three shifts per day, seven days per week, 52 weeks per year to prepare Columbia for her historic date with space history. I can remember spending ten hours on Saturdays and Sundays "pushing paper" in the Orbiter Processing Facility bay where Columbia was being processed. I called engineers on the phone, sat down with them, and recopied their instructions on Orbiter work documents so that we could get the pile of backlogged work completed. I often thought that there were millions of sane people across our land enjoying two days of rest and recreation while we slaved inside a hangar. But we all loved it! It challenged us! It demanded conquering! It put a red flag before us and presented a Matterhorn we were determined to climb! I thank God that I was sensible enough to cling to my wonderful family, to find time to attend early church services on Sunday morning and evening services on Sunday night. Not since the long, strenuous Apollo test operations had I experienced such a drain on the mind, body, and often the spirit.

Undoubtedly the most frustrating aspect of Columbia's first flow was the overwhelming amount of paperwork needing to be processed. The term "anomaly" has become a NASA byword delineating what NASA considers to be any situation that is not within a set of predefined limits. In the peak of the STS-1 flow, it was not uncommon to see Columbia's open paper well above 8000 in number. Most of this open paper was written to document anomalies.

Most of the anomalies were real. They would require anywhere from four hours to two weeks to be corrected and retested. These anomalies included broken wires, damaged tiles, leaking plumbing, failed black boxes (avionics units), and all manner of problems, all bad. Often when a technician went into an Orbiter compartment to repair a broken wiring segment, his access into confined areas caused him to damage two or three wires in an attempt to repair one. The Orbiter technicians came up with a strong, yet flexible, pallet material, which they dubbed "elephant hide". Those protective pads were used extensively to prevent wire and mechanical damage by workmen.

We started an all-out crusade to fix all the anomalies and clear the orbiter paperwork. We scheduled "paperwork parties" on weekends and holidays so that all our engineering resources could be leveled against the mountain of open paper. I recall the utter frustration of coming in on a Monday morning after a weekend paperwork party to find the total count of open paper to be greater than before the party! It seemed we often took two steps backward for each step forward. It caused some of us to begin to worry that the problem of open work was going to be insurmountable. Only strong determination prevailed and the open paper, along with the tile work, caused Columbia's lengthy flight delay.

Often in the midst of an unusually bad predicament, God reaches out through the hand of an angel to lift us up above the trials of life. In the midst of the Columbia paperwork woes, I was to be blessed by a simple request from my youngest angel daughter Wendy. It took the form of paperwork, sweet paperwork, not the kind we were battling 24 hours a day on Columbia. Often I would come home after dark when nine-year-old Wendy would have been tucked away fast asleep. One evening, I slipped in late, opened the freezer door to get ice, and found in the icemaker a beautifully composed piece of paper from Wendy. It was written on note paper that was stenciled

> **"CONFIDENTIAL, IMMEDIATE ATTENTION, EXPEDITE!"** "Mom, sometime this week, can I get an early birthday present? That dog in Plum Tree, it's only $30.00. If I can, that's all I want for my birthday,

> but going to Disneyworld. Ask Dad if I can get it, O.K.? I love you! Love, Wendy (P.S. I stuck some money in here for the dog!)"

Wendy knew Dad would find the note and two dollar bills in the icemaker. She can never know how much the touch of an angel's fingertips can make a multitude of headaches seem to totally disappear! Finding this new piece of paperwork to answer was a sure Godsend, one of the simple ways we are blessed in times of trouble.

The Kennedy test engineers have always been ultra conservative in their test philosophy. They are always complemented in this conservatism by their government counterparts, the subsystems managers at the Johnson Space Center, or the chief engineers at the Marshall Spaceflight Center. With this history, we obviously tested the Orbiter Columbia almost to the point of absurdity before its first flight. We actually performed a firing of the Orbiter/External Tank separation pyrotechnic bolts and ordnance in the Orbiter Processing Facility. We placed three 55 gallon containers of white Florida sand underneath the Orbiter's belly to absorb the shock and catch the single forward External Tank separation bolt and the two aft External Tank separation bolts. It was highly successful; no problems! We proved again that pyrotechnics will explode when induced with electrical current. As facetious as it may sound, a lot of us questioned the need for such detailed testing. Needless to say we never ran this test again after STS-1.

We also installed elaborate exhaust ducts in the ceiling of the Orbiter hangar so we could test fire the three Orbiter Auxiliary Power Units, or APUs. During this week or so of preparation and hot fire, an instance that I later found amusing occurred. Our Rockwell contractor's best test director was a jolly Florida Gator, Tal Webb. Tal and I had worked together through a lot of hard times getting space hardware ready to launch. We had set a Launch Commit Criteria (redline) limit of 500 parts per million of hydrazine allowed in the Orbiter aft compartment for the call to be made to shut down the three APUs. We in the engineering ranks were ready to call shutdown either at 500 ppm or 5%. Tal, the tried and trusted operations-minded test director, was not listening for percentage call-outs. He was expecting to hear hydrazine content in parts per million. About one minute or so into a very successful APU hot fire, the APU console engineer casually reported, as a matter of technical information, "The aft compartment hydrazine content is 2% and holding." Immediately, in an urgent voice, Tal called me to say he was going to order the APUs be shut down because of the hydrazine. "Are you sure you want to do that, Tal?" I asked, almost dumbfounded. Immediately Tal ordered the three APUs to be deactivated. As the participants began to discuss the hydrazine content, it soon became apparent to the test team that Tal's interpretation of the hydrazine percentages was out of redline when actually the hydrazine content in parts per million was certainly acceptable to continue. I could not resist the chance to speak up on the net and say, "Looks like we just shut down three good APUs." Tal and I often replayed that episode and laughed at what was a costly experience, but one in which we all learned some valuable lessons. To be quite honest, test firing three hazardous, hydrazine-filled power units was ridiculous. We held a lot of "science fairs," the title we give to overextended testing today.

We laughed at ourselves often as we pondered the reasons for all of this unnecessary and inappropriate testing. I remember a comical comment by Bo Bobko, one of the first Shuttle astronauts. Bo had been working closely with us in writing the numerous and complex countdown procedures needed to launch the first Shuttle. He had a keen working knowledge of the Orbiter and followed closely all the numerous tests we performed. One day Bo related his idea of how the Russians would view our test activities. We were all well aware that the Soviets were racing to get their Orbiter look-alike, the Buran, ready to fly. Bo in his best Russian accent gave us his idea of what a telephone conversation, coded of course, would sound like between two Russians, one a spy watching us prepare the Shuttle for its first flight; the other his leader back in the Soviet Union. Boris the spy is calling Ivan the big boss back home!

"Hello, Ivan, this is Boris! I'm ready to give my weekly update on the NASA Shuttle program. You won't believe what these crazy Americans did over the last few days!"

Boris proceeded to give Ivan a running account of the details of the KSC Shuttle testing day-by-day. His supervisor listened intently to the field report and when Boris had finished, Ivan could make only one terse comment!

"Boris, I'm afraid you've been away from Moscow for far too long. You are hallucinating! You are fabricating stories about those crazy Americans that I just can't believe! You better catch the very next flight home for a period of rest and relaxation. You have been in America way too long, Comrade Boris."

Bo often laughed as he recalled the instance when he was supporting extended testing on Columbia's rudder pedal assembly. The engineers on the voice net asked Bo to push downward on the rudder pedals with his feet and legs as they made measurements and collected systems data. Bo recalled how they seem to have used his lower extremities as ground support equipment for at least three of four hours. When he didn't hear them talking after thirty minutes or so, he called the test engineer but received no response. When he released the rudder pedal, no one raised an objection. Much to Bo's chagrin, he found that the engineers had deserted him and had gone to lunch.

KSC was "blessed" by having a noted project manager from the Johnson Space Center assigned to direct the Orbiter processing during its hangar time. The former NASA project man was given the formidable tasks of recovering the lost schedule time. This was too much for any one leader. In retrospect the gentleman whose name doesn't need to be mentioned probably made a lot more enemies during this time than friends. Such was the natural expectation from a job too big for any one set of britches.

One of the "imported" managers, as impressive to himself as to those who watched him perform, once told a small group from the OPF workforce that he certainly thought they were "assiduous." One of the macho tech supervisors almost needed to be physically constrained to avoid attacking the manager. With his limited vocabulary, he was certain the proud Orbiter team members had been verbally insulted and dishonored!

We were also provided the leadership of a test czar, Astronaut Bob Overmeyer, who we learned to call affectionately "Captain

America." I was assigned directly as a staff project engineer to help Bob test the Orbiter the right way. Again the magnitude of the battle often overcomes the best soldiers chosen to fight them. We learned to accept Captain America's "gallant speeches" and "lance and spear charges" as part of the job. Bob became a good pal and I was pleased to escort his crew around KSC in 1985 prior to his first Shuttle flight as commander. His crew wanted to personally thank thousands of Kennedy workers for their support. Tragically Bob Overmeyer was killed flying an experimental aircraft for a private company after his retirement from the astronaut corps.

As we reflected back on that first Orbiter flow through the test and operations cycle, I had two impressions - one mine and another suggested. I often thought how our huge test team of government and contractor personnel was actually much like a group of kindergarten youngsters asked to assemble one of those large 1000 piece jigsaw puzzles. I imagine we were as lost in putting together a new space vehicle piece by piece as they would be putting together the puzzle depicting a hillside snow scene in Vermont. As difficult a challenge as those white puzzle pieces would be to those kids was much how the intricate pieces of Orbiter hardware appeared to us.

Al McGee, one of our best potty engineers, suggested his idea of how the Orbiter flow seemed to progress. Al compared our efforts of getting the Orbiter ready to Sisyphus of Greek mythology. Sisyphus was destined to forever toil in Hades by pushing a huge stone up a steep hillside until he almost reached the top. Just as he was about to reach the summit, the heavy stone rolled over him and landed at the bottom of the hill. Time and again Sisyphus was required to push the stone up to the top only to have it roll back down again. No better analogy could be made of our Columbia first flow operations. We were definitely in a Sisyphean situation - or it sure felt that bad!

Despite the added management help and the hordes of tile technicians who came to Florida from the Palmdale, California Orbiter manufacturing plant, we managed to struggle through the long, long Orbiter processing period and arrived at the pad. I will always be amazed at the relatively easy pad stay for the first launch of a new vehicle. We only scrubbed one launch attempt when the set of four onboard computers failed to properly communicate prior to launch.

And even better than the excitement and exuberant joy of launching a new era of space exploration, we had the privilege of providing the ride into space and back in a reusable ship to two great space heroes, John Young and Bob Crippen.

It was during the Columbia preparations for the STS-1 mission that astronaut John Young again started writing his famous "John Young letters" pointing out concerns he had about certain aspects of the program, the hardware, procedures, or generally how decisions were being made. NASA managers and engineers all came to dread seeing another Young letter, but we also came to appreciate the value and candor of John's writings. This practice continued throughout the early Shuttle days. John's pencil kept the manned space program on its toes!

The accumulated tile work, the excessive amount of testing, and the thousands of open paperwork discrepancies against the hardware and its ground support equipment were horrendous problems to overcome prior to the first Shuttle launch. Even so, the most frustrating, the most challenging, and the most character building was the task of making the ground software and the flight software operate together. Again almost in a testing mania, we scheduled an integrated test in every facility we took the Orbiter or the Shuttle stack. We ran OIT, Orbiter Integrated Test; we ran SIT, Shuttle Integrated Test; we ran DIT, and I don't even recall what this acronym represented. Dynamic Integrated Test, I guess! We performed these tests several times. It seems that every time we made a software modification, either flight or ground, we were required to repeat the appropriate integrated test. Each integrated test required a firing room crew for days on end, 24 hours per day, seven days per week. The main players in this saga were the new ground launch sequencer, the GLS, and its cousin, the onboard redundant set launch sequencer, the RSLS.

While we were working day and night to make the flight and ground software play, the engineers at the Johnson Space Center were experiencing the same nightmares. They were constantly running the flight software through the SAIL, the Shuttle Avionics Integration Laboratory, a good acronym but by no means a description of how easy things moved through the facility. This hardware/software complement of computers and Shuttle avionics black boxes was an excellent test bed for "debugging" flight and ground software packages. It was interesting to learn that one of the key players during the STS-1 SAIL software certifications, long and arduous days of constant on-station time, was a young promising astronaut, Dick Scobee.

We would preshift enough Firing Room personnel to get the Orbiter and its systems ready for the integrated test on third shift. Then the prime launch team led by the NASA test director, the Shuttle Project Engineer, the NASA Orbiter Project Engineer, and the contractor Test Project Engineers, would get the integrated test started by simulating a time of about T-3 hours in the countdown. For 12-16 hours, daylight to dark, we struggled to overcome holds, data funnies, unexplained conditions, every hardware/software problem imaginable. Over and over we failed to complete a totally acceptable integrated run and dragged into the operations scheduling room at 6-8 o'clock in the evening to plan for the next attempt on the following day. Our "leader" or "big boss" was the Gorilla himself, George Page. Up to that time, George had been test director for the Gemini and Apollo programs and was by far the best pusher and operations expert in the program. He was also a strict disciplinarian. He took no "crap" as he often reminded us. It was his custom to be at our 6 p.m. debriefs to "chew us out" about how we better have a successful run to follow or "else." The two people receiving the brunt of his "candid wrath" were Bob Sieck, the NASA Shuttle PE, and me, the Orbiter PE. We happened to be the two engineers who led the engineering troops. George was the ultimate operations engineer. It was operations' job to constantly maintain pressure on engineering to get its testing and work done on time. Looking back on this era, I realize how control must be used to influence results under certain circumstances. I sometimes wonder how much damage was done to families and careers, and other intangible effects of such a grueling, mind-boggling experience.

After the success of STS-1, we all came to recognize that victory does not come easy. You often suffer casualties and the scars of battle remain forever.

A tragic automobile accident on Highway US 1 in Titusville claimed the life of one of Rockwell's outstanding engineers, Wayne

Fisher. Wayne was at the time leading the Rockwell support required to process Columbia, the first Orbiter, through KSC. He was a dedicated and knowledgeable individual who knew the particulars of the Orbiter long before it left Palmdale for KSC. I'm sure our long period of STS-1 would have been much easier had this "superstar" been with us. The death of "Fish" was a great loss to a lot of us who remember his big smile and his "can-do" attitude!

After completion of one of the long countdown simulations in February 1981, I was walking to my car in the VAB parking lot when I heard an emergency announcement and ambulances being scrambled to Pad A. I ran back to my office to learn that two technicians had been killed in Columbia's aft compartment. In haste to get work accomplished, confusion had arisen as to whether the nitrogen purge had been removed from the aft engine area. We purge this area in pre-launch with nitrogen to combat combustible oxygen and hydrogen leaks.

Two techs, the very best Rockwell employed, had been given work instruction paperwork to perform important pre-launch activities in the cramped portion of the Orbiter that is undoubtedly the most hazardous. Unsung heroes, men with expertise and dedication, John Bjornstad and Forrest Cole, were to give their lives in a tragic accident. The nitrogen purge had actually not been removed and the men entered a deadly environment. A 100% nitrogen atmosphere starves the human brain of oxygen in a little more than twenty seconds! These unsuspecting men died because we were pressing to make the first launch happen. I am saddened by that fact and it brought a sense of disappointment and sadness to the success we finally achieved on STS-1. My heart goes out to the families of these two outstanding men. NASA has not memorialized their sacrifice as we should have done!

I'm pleased that, to my knowledge, there were few divorces among the close-knit launch team personnel. Divorce and broken families had been almost prevalent in the early manned programs and the moon-landing era. I know we cemented relationships where engineers grew to trust and respect the other engineers that they depended upon. I recall very few temper flare-ups or turf battles. We were all pointed in the same direction; we had an ultimate goal in mind; and we had great leadership. I suppose the real driver to keep us fired up and charging was the thrill of a new technological challenge with opportunities to learn new systems and ways of doing space launch operations. And we loved it!

I could more aptly describe why we put so much work into the Shuttle program by borrowing from one of my favorite practitioners, Hagar the Horrible. In a classic comic strip episode, Hagar's wife had enjoyed a very good day and mused as she thought, "I wonder if something bad will happen tonight." Then she looked from the window and saw Hagar coming home from the battle of the day. She yelled to her son, "Here comes your father. Get the hot water and bandages ready!" She ran to the door to greet Hagar. "You poor dear! How did your day go?" Entering the sanctuary of home, bleeding, his nose big and red, a hatchet protruding from his helmet, three arrows stuck in his arm shield, and his head spinning with stars, Hagar replies, "Terrible! I lost the battle and the loot. I was beaten up and my boat sank!" "Well, you just sit here and I'll get you a drink and some warm food," his wife said. Bringing his dad a pot of hot tea, his son asked, "Gosh, dad, how can you take this year after year?" Hagar sat back proudly, pointing his finger to emphasize a point. "The secret is to like what you're doing, my son!"

CHAPTER 12

"We Have Main Engine Shutdown"

There are a few words and phrases that flight controllers, test conductors, and launch directors do not like to hear. "Abort" is a dreaded word to flight directors, whether during a simulation or God forbid, during an actual mission sequence. The launch team at the Kennedy Space Center always aims for a launch countdown where T-0 is reached without hearing the potentially perilous, "We have an RSLS Abort!" In the last 31 seconds of a Shuttle countdown, the onboard general purpose computers have control of the countdown events and work hand-in-hand with ground computers to launch the vehicle. These flight computers run a complex software program called the Redundant Set Launch Sequencer, or RSLS, onboard the Orbiter. The Ground Launch Sequencer, or GLS, is the equivalent ground software program that controls ground systems. This software communicates via a one million bits per second data bus with the onboard computers. If either the onboard software or the ground sequencer running in a computer at the Integration Console senses a serious hardware problem, the countdown is stopped. Inside the T-31 second time frame, a countdown stop or "hold" is considered to be major trouble and is handled as an abort situation. We experienced a lot of GLS aborts in the early days of Shuttle testing when we were trying to work the bugs out of the ground systems and the key software programs. But never did we experience a more white-knuckle, spine-tingling, hair-raising, and potentially tragic event in ground test history than the first main engine shut down on the pad in June 1984. We were very near launching Discovery on its maiden flight when this traumatic shut down hit us like a slap to the face.

After what seemed an eternity of emergency reactions to the pad shut down of the three main engines, we came out of the unfortunate incident very well. Only about forty minutes elapsed from the time of engine shutdown to the time we had safely egressed the astronaut crew from the extremely hazardous environment on Pad A.

We had spent many long hours studying aborts and contingency situations as we wrote the Shuttle test and checkout procedures. We simply referred to these as "what if" sessions. What if the SRBs don't ignite? What if one SRB ignites and the other one does not? We called this the "cart-wheel" scenario. What if the hold-down post pyrotechnics don't fire? What if the main engines shut down? What if we experience a significantly large hypergolic or cryogenic fluid leak in the last minutes of a countdown? And a hundred more "what ifs," some highly probable, some so unlikely to occur that we only discussed them and never gave their likelihood of occurrence any real credibility. So we were prepared for a main engine shut down on the pad. Or we sure thought we were! Even with the hours of procedure refinement and many failure response simulations behind us, many of us wondered if we were properly prepared for the real McCoy. I'm sure that infantry men down through history must have felt the same way when, after weeks of basic training, rifle range firing, and battleground war games, they heard the first rifle fire or felt the first blast from the enemy's artillery.

We had speculated that a shut down of the main engines on the pad was a scenario of relatively high probability. We had designed a system called the Firex water system to perform two important functions; provide a deluge of water to extinguish fires on the launch structure caused by the Shuttle's launch flames and to provide water at certain points on the pad to fight the probable fires following an engine abort. The onboard computer and main engine software provided the commands to sequentially and orderly shut down and safe the main engine systems. We were relying on the engineers at the fluids consoles in the Firing Room to manually command the ground deluge water flow if needed after an engine abort. We learned a lesson indeed during the first engine abort. We learned not to leave water activation to the discretion of an operator, to let the ground software command it on automatically. We learned that water deluge activation should always occur. We should not wait for engineering judgment to determine the need for water. Let it flow!

The *Aviation Week and Space Technology* edition of July 2, 1984 devoted a long article to our "heroism" in handling the June 26 engine shut down on Pad A. My job was Shuttle Project Engineer. I was in charge of the technical aspects of the countdown. I was the firing room leader of the NASA and contractor engineers on the launch team. Never before in the Shuttle program had ground controllers been subjected to an extremely hazardous situation with astronauts involved. Norm Carlson, an Apollo test director long on experience and "moxie," was the Chief NASA Test Director and my counterpart for operations. I always felt a little better when Norm was an involved participant in any launch countdown.

Everything about Discovery's launch countdown had been progressing well on June 26, 1984 until Main Engine 3 enunciated a failure identification, shortened to failure ID, or FID. A failure ID at T-2 seconds is very close to T-0. After startup, engine 3 sent a failure command which actually never provided engines 1 and 2 a chance to reach full power before their start sequences were interrupted. The confusion caused by an engine shutdown scenario we had not expected created a lot of other disorder trying to decide what reactions we should take.

I felt for the first time the hard cold reality that we were in an extremely hazardous environment with five crew members in the Orbiter sitting in a highly dangerous situation. I realized that two or three of us either in the launch firing room or the backup firing room must assume a strong leadership role. After advice from the fluids console and other key players, I was convinced we had more than one hydrogen fire around the Orbiter tail section and engine area. Hydrogen is one of the most volatile high explosive gases in nature. A hydrogen explosion annihilates most everything in its path. We expected hydrogen and oxygen exhausts after engine shutdown and we got both this first time. In my mind, I was concerned that water on the highly sensitive thermal protection tiles and blankets would ruin their thermal resistance capabilities. I also knew a hydrogen fire of close to 600 degrees F could overheat the tanks on the Shuttle and produce an explosion. The only overriding thought in my mind was my concern for the safety of the crew and their welfare. In times of chaos, when human lives are at stake, I think we ignore hardware, money, and other similar considerations. I felt we needed to be sure we had the hydrogen fires under control before we called for the crew to perform a Mode 1 abort. In this abort scenario, the astronauts are to exit the Orbiter unaided by the ground crews and slide to a safety bunker in metal baskets attached to a slide wire system. After advising Test Director Norm Carlson to let the crew remain onboard longer, I began to call for the responsible ground controllers to turn the deluge water on the vehicle and the structure. It took far too long to get the water activated, but we finally accomplished the task. I'll never know for sure how much of that long wait was caused by an engineer's reluctance to respond because he was concerned about water damage to flight hardware. Most of us processing the Shuttle were super sensitive to the Orbiter tiles and their susceptibility to water damage. We finally drenched the engine boat-tail area pretty thoroughly with water for nearly five minutes, turning the water on and off several times when the persistent hydrogen fires flared up again and again.

We were fortunate that day to see the STS-41D crew finally scramble and egress the pad structure safe and sound. Among the crew members, including my good friend Commander Hank Hartsfield, was a brilliant young woman Judy Resnick. I shall always remember watching Judy on the console's color TV monitor as she literally ran out the small Orbiter side hatch apparently without even touching the edges. Over three hours earlier, carefully and meticulously she had slid through the same oval hatch opening into her seat. As we all know, under pressure, we often make the impossible things look easy.

We won several accolades for our handling of the first main engine abort. I could not possibly single out the heroes in the firing room that day. Don Weinburg, Orbiter Test Conductor, was cool and performed well as did Al DeLuna, the Orbiter Test Project Engineer, and Nora Lavinka, the GLS operator. The launch team handled the hazardous abort like professionals. Never did anyone project a feeling of fear or panic. I was proud to be a part of this extremely successful accomplishment.

As the smoke cleared in the next few weeks, I realized how unpredictable a Shuttle countdown could become. I postulated that there were upwards of 1000 failure situations that could occur and cause you to have a real bad day. I also felt the enormous responsibility that leaders assume in similar positions. I realized that an airline pilot who faces a perilous failure in some phase of a flight is under extreme pressure to save the lives of his passengers. A platoon leader commanding a group of soldiers in face-to-face combat has a lot of pressure to protect the lives of his men. I soon came to understand that we must have the best training, test discipline, and tried and true procedures to be a top notch launch team. From that time forward, as I advanced in responsibility from Shuttle Project Engineer to Launch Director, I constantly emphasized to everyone involved that the safety of the astronauts was our number one priority. We were not to compromise anything to the detriment of crew safety. Having this commitment as a banner under which I led the launch team as its Launch Director, I was almost brought to tears to read in a post-Challenger documentary how Gene Thomas was not as cautious as he should have been when he launched Challenger. My career was dedicated to the success of the manned space program, but of primary importance was the safety of the men and women of our astronaut corps.

We drastically changed our test procedures and redesigned several pad systems due to the lessons we learned. Today an engine shutdown inside T-6.6 seconds, when the first engine start command is sent, is handled so routinely that it sounds like a normal day at work. The launch team no longer spends weeks recovering from an engine abort on the pad. Though never routine and never welcomed, an engine shutdown is no longer a dark cloud hanging over our heads.

One of the most refreshing notes received after this main engine shutdown incident was a telegram sent by folk singer John Denver, who was at the Kennedy Space Center for this aborted launch attempt, to NASA Administrator James Beggs.

"Dear Mr. Beggs, Congratulations to you, the entire launch team, and the crew of the Space Shuttle Discovery, on handling a most difficult situation with great courage and professionalism. You honor all of us as Americans and as human beings. I am proud to feel even a small part of our space effort and greatly look forward to the opportunity of working more closely with you. With great respect and love, John Denver."

Only two short years later, the effervescent balladeer was to pay tribute to the Challenger crew through the lyrics of one of his songs:

> *Yesterday I had a dream about dying,*
> *About laying to rest and then flying,*
> *How the moment at hand is the only thing we know,*
> *And I lay in my bed and I wonder,*
>
> *After all has been said and is done for,*
> *Why is it thus we are here and so soon we are gone?"*

Sadly, John Denver was killed in the crash of a newly acquired experimental-class aircraft on October 12, 1997. Denver was doing what he loved so much, piloting a small airplane along the coast of Monterey, California. In Denver's mind, flying allowed him to "float through the air like an eagle and look at the mountains as only eagles can."

> *"They fly away like an eagle toward heaven."*
>
> Proverbs 23:5

CHAPTER 13

The STS-61C Launch Fiasco

My old standby reference, the Webster New World Dictionary of 1957, defines "fiasco" as a complete failure or action that comes to a ridiculous end. In the true sense of the word, STS-61C was not a fiasco since we were eventually successful. Better described, it was a string of fiascos. I hesitate to refer to the long, stressful STS-61C mission scrubs and delays as failures. Yet in the accepted vernacular of today's world, they were! We actually experienced six launch delays, four of which were scrubs, after having ingressed the Columbia's flight crew. We started the Mission 61C launch attempts in mid-December 1985. The first scheduled launch date on December 18, 1985 was delayed one day due to additional time needed to close out the Orbiter's aft compartment. The final launch day of January 12, 1986 gave all of the Kennedy Space Center and NASA a lot to rejoice about.

There was a tremendous amount of activity at the Kennedy Space Center in the late 1985 and early 1986 timeframe. We were modifying one of the launch pads to provide a "roll-away" umbilical to service the Centaur stage with highly volatile liquid hydrogen. We were busy preparing the Orbiter Discovery for delivery to Vandenberg Air Force Base in California. Since 1979, the Air Force with NASA involvement had been constructing a new 2.8 billion dollar West Coast military Shuttle launch complex. My good friend Bob Crippen was to command the first launch from the West Coast pad and one of his crew was to be Pete Aldrich, secretary of the Air Force at the time. Secretary Aldrich was following the lead of Senator Garn and Congressman Nelson in flying the Shuttle. He felt it appropriate that he should fly on the first Air Force Shuttle mission flown from Vandenberg. I spent a lot of days at Vandenberg on temporary duty as a Shuttle Project Engineer in 1983-84. I was scheduled to serve as Launch Director for the first several Vandenberg launches until the Air Force officers were ready to assume the major launch control responsibilities.

Newspapers were writing about the hectic pace at which we were working at Kennedy. In September 1985 the Denver Post reported "Not since the days of the Apollo space program have manned launch preparations rushed along at such a feverish pace. With one Shuttle on the pad and three waiting in the wings, the United States is poised to send crews into space with an airline schedule regularity, undreamed of a generation ago - a flight every 3 1/2 weeks next year."

We were trying extremely hard to avoid unnecessary duplications in our test program. We were working to reduce the time required to integrate a mission's cargo into the flight production process from 18 months before launch to 7 1/2 months. This meant faster generation of the engineering and operations products that were required for a given mission.

Tom Utsman, the Kennedy Deputy Director, honestly admitted to the press that "we are probably at our peak workload as far as the Shuttle goes. Some of our people are working as hard as they've ever worked." Tom was a real watchdog, determined that the launch team would remain alert, efficient, and, most importantly of all, safe.

Under Administrator James Beggs' signature, NASA had just responded to the 1984 annual safety review of NASA by the Aerospace Safety Advisory Panel. The panel had not been expressly critical of NASA's safety programs, but, as they normally do, they brought the emphasis again to the safety of the crew and flight vehicle. I found especially sensitive language in one of NASA's Headquarters' response to the panel's recommendation 1.5. The panel recommendation states: "NASA management would be well advised to avoid advertising the Shuttle as being operational in the airline sense when it clearly isn't. More to the point, however, is the fact that Shuttle operations for the next five to ten years are not likely to achieve the routine character associated with commercial airline operations".

I was quite amazed to read NASA's answer to ASAP's 1985 recommendation 1.5: "NASA's highest priority is to make the Nation's Space Transportation System (STS) fully operational and cost-effective in providing routine access to space." This priority declaration by the Administrator to the top safety advisory panel may have been strongly influenced by political interests. The House Science and Technology Committee report on the Challenger accident actually recognized that the Committee itself, the Congress and the Administration had played a contributing role in creating the pressure on NASA to achieve a launch schedule of 24 flights per year. They also candidly reported that Congressional and Administrative policy and posture indicated that a reliable flight schedule with internationally competitive flight costs was a near-term objective.

I do not know to what depth in the NASA organizational structures this pressure was applied. I do not know what level of management at the Kennedy Space Center felt any pressure from outside to accelerate the schedule in any way. Whatever and wherever there was launch pressure, I never once felt it from above or from outside the Center. On the contrary, very few days went by that Tom Utsman or Bob Sieck did not solicit my evaluation of whether we were trying to do too much, compromising safety, or asking our employees to put in too much time at their jobs. We practiced safety. We held safety as our top priority, not 24 flights per year!

Nevertheless, we were again pushing as much as we were during the STS-1 processing days. It was a fast pace, but a safe pace. There was no doubt in my mind that the hard charging Shuttle launch team would get the job done and meet the schedule challenges. I also knew and trusted each of them enough to know that they would quickly blow the whistle if we pressed too hard. These engineers, attuned to pressure in the launch environment, were by no means dummies and certainly were outspoken. I never knew an engineer or technician to be shy about letting you know his position on a technical problem or an operational issue. We also empowered each of the engineering disciplines, or systems, to the extent that if they had strong concerns about the work load or the employees' physical or mental conditions, they could actually refuse to continue until corrective action was taken.

It was in this atmosphere that we transitioned into the Mission STS-61C fiasco sequence. We had a tremendously popular crew commanded by Hoot Gibson. Well-loved Charlie Bolden was Columbia's 61C pilot. The crew members included Dr. Steve Hawley and our local Florida Congressman Bill Nelson. There had been rumbles made by the press that Greg Jarvis had been bumped from the STS-61C mission to allow Congressman Nelson to fly late in 1985. In my position as the Launch Director, I was in no way involved in the selection of the crew or the cargo to be flown. As Director of Launch and Landing Operations, my engineers and I were responsible to review the proposed launch dates, evaluate our resources and workload, and agree to the flight manifest, nothing more than an integrated launch schedule. I do know that manifest changes were plentiful and any reassignment of crew members from one flight to another or cargo deferral from one flight to another was considered to be almost routine by our Kennedy manifest support planning team.

We scrubbed the December 19, 1985 Columbia launch attempt inside T-31 seconds when one of the right hand solid rocket booster hydraulic power units failed to reach required turbine speed in the proper manner in the specified amount of time. A Criticality I hardware item, the power unit must be operational to allow gimbaling of the booster's nozzle during ascent burn. This was a distinct hardware failure and the Ground Launch Sequencer did its job and stopped the countdown. Due to the involved work in removing and replacing the power unit in the booster aft skirt and the tests required to recertify the hydraulic system, we smartly decided to delay 61C over the Christmas holidays. We set a new launch date of January 6, 1986. This became the first of a long string of disappointments in launching Columbia's 61-C flight.

On January 6, we almost made a crucial mistake after a launch abort at T-31 seconds due to a liquid oxygen fill and drain valve problem. After a lot of last second attempts at getting the main propulsion system configured properly, we exceeded the engine cool-down temperatures required to start the three main engines. After recycling the countdown clock to T-20 minutes, we held at T-9 minutes to meet our propulsion system redlines, and unavoidably exceeded the SATCOM payload's launch window for that day. Post test evaluation showed that the LOX console had actually drained up to 18,000 pounds of liquid oxygen during the initial hold. The LOX systems engineers had never experienced a situation such as this and inadvertently overlooked this critical condition. Had we proceeded with the launch that day with less than enough liquid oxygen, we could have possibly failed to make orbit because of negative margins. After discussing this anomalous situation post-launch, I realized for the first time that I had relied on systems experts who had given a "go" for launch and the system had broken down. A Launch Director cannot possibly know all the intricate details of the Shuttle systems, not even the critical systems. A team of LOX and main propulsion experts from all over the country had failed to realize that we had less than enough of one of our two main fuels onboard. No one had enunciated a countdown hold. There was an established launch abort process. This process had to work. This time it failed us. By the grace of God, other circumstances prevented STS-61C's launch on January 6, 1986. We learned a significant lesson from this costly oversight. It made every member of the launch team more aware of the insidious failures that can bring you to your knees. We were better from every lesson we learned! It also made a strong impression on my responsibilities as Launch Director. I saw so vividly demonstrated how the fate of a crew or mission can hang precariously by a single thread! If the string snaps, you court disaster. If the chain of events and conditions are favorable, you dodge a bullet and avoid a catastrophic outcome. Often a judgment call, although based on engineering experience and technical know-how by one or two key individuals, can adversely affect the destinies of thousands of others. Such was to be the case later on in the month when Challenger and its crew were lost.

On our January 7, 1986 attempt to launch Columbia, the dreaded weather monster began to attack us. Following a rather uneventful countdown, we ingressed the crew again. We shortly began to get a "no go" from the Houston Flight Director and weather officers because of adverse weather at the Transatlantic Abort Landing (TAL) sites in the Southwestern Europe-northwestern Africa emergency landing regions. Our Launch Commit Criteria dictated that we must have an acceptable TAL site and it was not going to happen that day. We decided to recycle the clocks to T-20 minutes after progressing to T-9 minutes. We were able to get the mission planners and payload customers to extend our launch window for one hour. We held again at T-9 minutes on the second attempt to get to T-0. The weather at the TAL sites never cooperated and we went into a 48 hour scrub scenario. During the turnaround, we discovered a faulty temperature probe in the liquid oxygen system and our recovery plan for this problem drove us to delay the next launch attempt until January 10. It was a frustrating time for both crews, the flight crew and the launch team. I was beginning to think this particular mission would never succeed. The 61C crew was such a bunch of tremendously nice people that in my sense we never had a loss of composure or a criticism of one another's contribution. I feel the crews demonstrated a level of professionalism seen only in some of the other high-tech operations of our military services.

During this time, I came to respect the strong leadership of Hoot Gibson. I also began a long and lasting friendship with sparkling, smiling Charlie Bolden that I will hold dearly for eternity. I came to realize that a brilliant person like Dr. Steve Hawley could be describing celestial phenomena one moment and doing a comedy act in the next moment that would leave you in stitches. When we finally launched 61C on January 12, Steve came into the Orbiter white room to be readied to go through the Orbiter hatch. As we all watched the color TV monitors, Steve turned to face the camera wearing a Groucho Marx nose and the classic horn-rimmed glasses. He motioned as Groucho has done so often to insinuate that he was flipping the ash off a cigar with his index finger. I thought to myself in amusement, "What kind of circus am I the ringmaster of, anyway?" It really proved extremely beneficial to a launch team that had suffered through a lot of recent setbacks. I could see the smiles on the faces of the console operators as I scanned the Firing

Room. Steve explained later that the crew had been ribbing him as being their jinx. Steve, along with Judy Resnick, had been one of the members of Discovery crew when we experienced the highly hazardous main engine abort on STS-41D. He held the record for the most scrubs experienced by a Shuttle crew member. Steve went on to explain how he figured if he adequately disguised himself, he might fool the Orbiter into thinking it was Marx flying instead of Hawley. That might relieve the jinx and allow us to be successful. As all highly engineered and technological breakthroughs like this, it worked well. We launched!

Some would tend to criticize well-intended instances like this. In reality, it had a tension breaking result that was superb. Psychologists recognize laughter and a good sense of humor as essential to an individual's well-being. Such was the breath of fresh air this simple prank brought to the Shuttle launch team. I thanked Steve several times for his astute play-acting.

In the string of STS-61C scrubs, I felt the launch team was struggling, but still up to the task. We allowed plenty of time between scrubs to refresh our bodies and minds. I did find that it became almost commonplace to get home late in the afternoon after another 61C scrub, tune in to the 6:00 television news to hear my voice talking to Commander Gibson, "Sorry, Columbia, but the weather just isn't going to let us get the launch off today. We'll give it another try tomorrow. Thanks."

Always, every time, Hoot came back calm and gentlemanly. He came across as totally understanding of our dilemma. "That's fine, Gene. We understand. We'll be ready to try again when you folks are. Thanks for the great support." This is what made every man and woman on the ground love and respect every man and woman who flew the Shuttle. This is why not a single one of us who never fly into space would in any manner jeopardize the precious lives of those who do fly into space.

During the 61C launch period, I also came to admire and respect Congressman Bill Nelson, now senior ranking Senator for the State of Florida. Like outspoken, professing Christians often do, Bill Nelson took a lot of criticism from a lot of factions. Many inside NASA resented Bill's free trip into space. Luckily, I never had to debate any of my associates or the press on the merits of flying politicians on the Shuttle. Bill did not deserve criticism. He approached the flight as a professional and trained strenuously to carry out a significant portion of the mission workload.

Bill's grandparents had actually homesteaded over 100 acres of this plush Merritt Island land that was now the Kennedy Space Center. As a boy, he had played in the woods around the area. Now his boyhood playground became a launch facility for him as a man. Serving his fourth term as a U.S. Representative, Bill Nelson was trying to assume two significant roles: to fly the Shuttle into space while trying to maintain some cognizance of and participation in the activities of the United States Congress.

Whether NASA asked Senator Garn or Congressman Nelson to come fly the Shuttle for political reasons will probably never be totally verified. After his Columbia flight, Nelson was probably the most supportive Congressman NASA has ever had.

The launch attempt on January 10 was not something that I reflect on with pride. For the first and only time in my long career in engineering and as the Shuttle Launch Director, the decision to go ahead with tanking, loading, and crew ingress for a launch attempt was made purely for political reasons. It is the sole instance when I have consciously been involved in a political shenanigan. I was aware that it was for some unknown cause or to make an impression somewhere. "Acting Administrator Graham is here for the launch and 'they' want to go ahead and take a shot at it." These are surely not the exact words spoken to me and I honestly don't recall which of the people above me in NASA's hierarchy spoke them. I don't think it was a Kennedy manager who I reported to, rather one of the Level II or Level I managers.

The weather could not be adequately described for that January 10, 1986 Florida day! Horrendous, unbelievable, foul, all these words fall short. The rain fell in torrents from late in the tanking phase through the entire day. I suggested we hold the crew at the crew quarters, but was again told that "they" want us to continue the countdown. At the time the crew passed the Launch Control Center, I turned to observe Pad A, their launch point. The rain was so thick that there appeared to be no Pad A or Pad B anymore. Vision was obscured for anything past 500 yards or so outside the launch center. Candidly speaking, I felt like a dunce to be ingressing a crew in a pouring rain in such adverse weather. The predicted weather was expected to be terrible with torrential rain and lightning well past our launch window. I played the game, always the good soldier. To this day, I consider that unnecessary expenditure of time and people to be ludicrous, a waste of taxpayer's monies. It was political, I am certain. However, I am just as certain that I never again, on my word of Christian ethics, ever participated in or have been motivated by pressure applied for political purposes. Nothing associated with the Challenger accident that I was involved with had any political drivers causing us to do something wrong. There are often pressures brought on the launch team. However, except for this rainy day countdown on Columbia, I have never been cognizant of a political agenda influencing my conduct. Grace be to God, the final time that I boldly declared we were going to scrub, no one objected. Holding at T-9 minutes, the crew's vision of the launch tower practically obscured by rain, lightning all over the area, I again informed Hoot that "there is no way the weather will clear today and we are scrubbing again." I think the congenial Commander replied something like, "Thanks, Gene, we knew that." The heavy rains had taken that launch opportunity from us and we decided to try again on Sunday, January 12.

All my life I had never considered luck to be real. What some refer to as luck I believe is more correctly good timing, opportunity, personal readiness, boldness, and probably a dozen or more similar attributes which an individual can possess or exhibit. One strong attribute that people often refer to as luck is to the Christian merely the grace of God. Anything in my life experience that someone might call luck, I strongly believe is, in reality, God working His will on my behalf. After the long rainy January 10 day and the eventual scrub, big Rex Hands, the lead electrical power systems engineer, approached the Launch Director's console with a big smile on his face. "Gene, I'm not sure you believe in luck, but maybe this will help us launch next time." He held out his huge open hand to give me a neat gray rabbit's foot that was probably intended to decorate someone's keychain. "Rex, I never count on luck in the equation, but I'll try anything. It certainly can't hurt. Thanks, big guy!" Today, the rabbit's foot is among the hundreds of momentos that I have collected at various points in my career.

A big cartoon in the *Tampa Tribune* later in the year depicted a very active Shuttle firing room with the Launch Director going over his final "Go-No Go" checklist with the launch team; "Fingers crossed? Check! Four leaf clover? Check! Rabbit's foot? Check!" After the slate of launch countdown problems, even the media began to wonder if we could use some pure luck.

Sunday morning was a charm! Temperatures were in the low 50s. Clear, crisp, gray Florida predawn gave way to clear azure skies. Calm winds and all TAL sites were "go," if I remember correctly. A perfect launch day was something the flight and ground crews sorely needed. After the Shuttle crews are into their seats and ready to launch, the Launch Director gives his final best wishes to the crew in the 10 minute hold at T-9 minutes. Recognizing how long we had toiled with the 61C crew I said, "Well, Columbia, looks like we're go today! I want to wish our best to the best crew we've had around here for a long time!" We had truly kept those guys around KSC for a long time. I know Hoot and the crew acknowledged my wordplay, but I'm not sure it didn't slip by a lot of listeners. Nevertheless, the January 12, 1986 launch was on time and led to an almost flawless mission. A week later the KSC newspaper, *Spaceport News,* was front page with Columbia's 6:55 a.m. January 12, 1986 launch. Bob Sieck's comments were "The launch team is tired, but very proud and happy." In the news conference after launch, I was reported to have said, "I don't remember the launch team demonstrating such clapping and exuberance since STS-1." I also went on to say that rounds of applause also followed booster separation and negative return or "No RTLS." I'm not sure what I could have said. I'm sure we must have been in a daze. I do distinctly remember the feeling of relief and gladness that we finally launched. And never do you forget the feeling of pride, pride in America's role as the leader of the world's space activities.

And what did we have to look forward to on Monday morning, January 13, 1986? The Shuttle launch team met in Firing Room 3, the other prime Firing Room, to run the Terminal Countdown

Demonstration Test (TCDT) for the Challenger crew led by Dick Scobee. Back-to-back significant tests on two Shuttle stacks, two flight crews, two firing rooms, two launch pads, one Launch Director, one NASA Test Director, one Shuttle Project Engineer, one Shuttle Launch Team! Such was the hustle and bustle leading to the fateful day in late January, 1986 when a very dark event in our history occurred.

I would by no means want to "fan the flames" that attributed Challenger's accident to launch schedule pressure. Nor would I ever try to downplay the tremendous amount of work we were doing; back-to-back launches, repeated scrubs, significant hardware modifications, and Vandenberg launch preparations. However, I will always and forever maintain that the key participants in the Shuttle launch process prior to Challenger were dedicated to the safety of our astronaut crews. I am persuaded that the men and women who comprise this elite flying corps knew that the Kennedy launch team kept safety as its number one priority.

CHAPTER 14

Setting the Stage

It is He who sits above the circle of the earth…

Isaiah 40:22

As Launch Director, I had been involved in five successful Shuttle launches prior to the Challenger accident. My first in that position, STS-51J, was launched on October 3, 1985 commanded by my good pal, Karol (Bo) Bobko and pilot Ron Grabe. This mission of Atlantis, a highly secretive DOD mission, lifted off at the exact time planned and was launched on a perfectly clear Florida day. Because of the secrecy of the mission, I had no responsibility to hold press conferences and felt no obligation to divulge any aspects of the countdown or the launch preparations. With perfect weather, a smooth countdown and an on-time liftoff, good friends as crewmen, and no press conferences, I laughingly told several of the launch team members, "This Launch Director job is a piece of cake."

In late October 1985, I had the privilege to launch STS 61-A, Spacelab D-1 commanded by the popular Auburn War Eagle Hank Hartsfield. Hank later was to play increasingly important roles in NASA's future. Our paths crossed several times as we both pursued other agency jobs and assignments. Hank's Spacelab crew was certainly international with two German scientists and one from the Netherlands. This flight of Challenger was also a pleasure and some of the news reporters actually wrote about how routine Shuttle launches had become since we had launched twice in the month of October 1985. This October 30 Challenger launch was witnessed by Christa McAuliffe, a charming school teacher scheduled to fly in 1986. Reports were that Christa was quite impressed with the smooth Shuttle operations that day.

The next launch I directed was very near Thanksgiving, November 26, 1985. The STS-61B crew was commanded by a tremendous gentleman and wonderful friend, Brewster Shaw. When first meeting Brewster during the STS-9 preparations, I thought "Don't tell me this scrawny little guy is going to fly Shuttle." What a foolish thought that turned out to be! Brewster Shaw became one of the very best commanders to fly an Orbiter and later held high level NASA management positions after Challenger. Few people know that the great American space hero Brewster Shaw was once a long-haired guitarist in a rock band in his younger days. Another gentleman who commanded respect flew right seat with Brewster. Bryan O'Connor, highly intelligent, a quiet Marine, also became a key astronaut who rose to leadership positions within the agency. Among the STS-61B crew was someone I remember well. Jerry Ross became an expert in extravehicular activities, and in 1996 I had the pleasure to appear with Jerry in a made-for-television video about Christians in the space program. "Circle of the Earth", from a Bible verse Charlie Duke remembered when he first looked at the earth while standing on the moon, was the title chosen by the producers of this ABC religious documentary.

The STS-61C Columbia crew that finally launched on January 12, 1986 after six launch scrubs was one that I especially admired and was close to. Someone might contend that when you start to attempt a launch on December 14, 1985 and experience launch scrubs until January 12, 1986, you have to get to know the key players. Commander Hoot Gibson was a pilot extraordinaire. Pilot Charlie Bolden became a dear and cherished friend over the years. Dr. Steve Hawley, a brilliant astrophysicist, is a very funny individual, a top notch manager, and one of the best long-armed shortstops to ever step on an amateur softball field. Franklin Chang-Diaz was a pleasure to know, a bright engineer, always smiling, one who loves to tinker. Pinky Nelson has so many hours in extravehicular activities that it is home to him. Congressman Bill Nelson, a strong Christian, became a close comrade. It was a special honor to launch the representative from your local district into orbit. Many citizens might jump at the chance to send their local politicians for a long journey in space! Almost daily, as we experienced scrub after scrub, I received calls from Bill in Washington checking on how we were progressing and how the weather was expected to behave. I was especially honored to finally and successfully launch this crew of super people. It is my strong professional opinion that we performed the long scrub-ridden STS-61C launch attempts in a safe manner. The post-Challenger accusations about lax safety at KSC in launch operations are unfounded when you look at our previous launch records. The dedicated men and women of NASA and its contractors were more deserving!

The launch team was operating on a high rate launch schedule, but I felt that we were never more ready. I could actually feel the spirit as we pulled together as a team, enjoying the push and pull that some of us called "recreational stress." After STS-51J, the initial flight of the Orbiter Atlantis, we found ourselves for the first time having four flight-worthy Orbiters in flow, Columbia, Challenger, Discovery, and the new fleet member Atlantis. There was a tremendous amount of activity. Often we went directly from one operation in one firing room on one stack to another operation in another firing room on yet another vehicle. We actually launched Columbia, STS-61C, on January 12, 1986 and manned the other firing room on January 13, 1986 to perform a Countdown Demonstration Test with Dick Scobee and his Challenger crew. Back-to-back major tests on two different Shuttle vehicles!

Most of our tried-and-true space veterans thrived on the activity. Tom Utsman, our former Shuttle Management and Operations boss, asked me at least once a week about the mood and condition of the launch team. I assured him that if I ever sensed a breakdown in any critical aspect of our job, any indication that safety was being compromised, I would call a "time-out" and regroup. I was totally confident that if I felt a need to slow down our operations, I would have the support of all the management above me.

Sure, we were tired! But we loved our jobs so much, we relished the feeling. There were some of us who were a little skeptical of why we were pressing so hard to modify a Shuttle launch pad and one of the Orbiters to accommodate a Centaur upper stage in the payload bay. We were scheduled to fly that Centaur stage in late summer of 1986. It was one of the few missions out ahead of us that on more than one occasion gave me a feeling that some of our designs and goals were pushing the envelope. Some of us were somewhat skeptical at the prospect of loading highly flammable liquid hydrogen into the Orbiter cargo bay to provide fuel for the Centaur stage. My few dealings with hydrogen or threats of its explosion were all bad. As a young engineer on the Gemini program in March 1965, I had watched a Centaur stage explode on Atlas Pad 36A from a short safe distance at Pad 19. The intense heat and consuming thermal energy released in that quick orange fireball made a lasting impression of the extreme hazard associated with hydrogen!

The Committee on Science and Technology of the U. S. House of Representatives was highly critical of the Shuttle schedule pressure in its October 19, 1986 report on its Investigation of the Challenger Accident. "The Committee found that NASA's drive to achieve a launch schedule of 24 flights per year created pressure throughout the agency that directly contributed to unsafe launch operations."

Of course, this House report was released after and because of the Challenger accident. It strongly infers that our launch operations were unsafe. The report goes on to later state that "the Launch Director failed to place safety paramount in evaluating the launch readiness of STS-51L."

These are strong accusations directed toward an agency that is so large and has such a gigantic technological operations base. This could be said of any similar large organization. There will often be a tendency not to prioritize safety by some involved parties if their existence depends upon meeting a schedule or accomplishing a specific task. We certainly were in a hurry at KSC! No denying that fact! We had safety concerns, as previously discussed, regarding the Centaur hydrogen-fueled rocket in the cargo bay. But that plan was not a KSC-originated change; the flight manifest and payload requirements drove the Centaur design. It is almost unimaginable to believe that anyone could review the launch attempts prior to Challenger, especially the six launch scrubs for STS-61C, to see that we did not have launch fever. We were not trigger happy! From the Launch Director down to the lowest level on the launch team, we launched when we felt we were safe to launch! We were not pressured by political agendas or threats to our budget! Those of us who were close to the day-to-day operations and responsible participants on the launch team were simply not exposed to politicians, bureaucrats, or Headquarters personnel who would have a reason to try to influence our schedules. As Launch Director, my policy was to talk to the media whenever appropriate. I never sought press and television coverage as I felt too much exposure only took your attention away from your job priorities. I appreciated the great corps of reporters, mostly younger men and women, whose prime beat was space. They were smart, not subject to be tricked or misled, and some of them knew the Shuttle systems as well as we did. It was a pleasure to work with these dedicated people, all fans of the space program and NASA. Not until after the Challenger accident did I recognize the insidious nature of numerous investigators and journalists; a lot of "outside" writers who could publish articles and books with any amount of hearsay or personal opinions. I never felt bitter toward any of these people, but I was often dismayed at the point to which some writers had digressed to get the public's attention or to simply "get published."

Such was the scene in late 1985 at the KSC. Four Orbiters, a busy manifest, Centaur flight preparations in work! I was enjoying my new job as Launch Director and Director of Launch and Landing Operations immensely. I had outstanding engineers and managers assigned to each flow and located at each major test and checkout facility. I had an excellent manifest guru, Gene Sestile, who really did the critical manifest scheduling for NASA's entire Shuttle program. I had a crackerjack Flow Director assigned to each Orbiter, someone who followed the Orbiter wherever it went, except into orbit. They were the men and women I leaned on to manage every aspect of the Orbiter's ground processing. They were the best: Ann Montgomery, Tip Talone (affectionately called "NASA's oldest teenager"), Conrad Nagel, and Jim Harrington, the Challenger Flow Director. And in concert with my Flow Directors, the NASA engineering director assigned four outstanding project engineers... one to each Orbiter. They were the engineering counterparts to the four Flow Directors. What a great team! The contractor was totally involved with one of their best operations engineers assigned to each Orbiter as Flow Managers to assist the NASA Flow Directors. Times were good; the reward for the long hours and hard work was a beautiful, heart-pounding Shuttle ascent through the Florida sky. I have never once heard a space worker comment that launches get boring. Each launch still thrills even the old space veteran and brings tears to the eyes of Americans who still believe in man's quest to explore.

From a personal, professional viewpoint, I felt I was on top of the world! My family was happy and our Christian walk was close. All three of our super children were in or near college age. There was never enough time for activities away from the hectic space program agenda, but our lives were full. I felt certain that I had the most exciting and, by far, the most rewarding job in all of America. God had been so good and blessed us immeasurably.

CHAPTER 15

The People Involved

One of the most difficult tasks that I've ever attempted was to answer queries about someone who had listed me as a reference on a resume or a job application. I find it is just as difficult to describe accurately the personalities of the numerous engineers and managers who were key players in the events prior to and surrounding the Challenger accident.

William Graham was the acting NASA Administrator, a political appointee of the Reagan Administration, reportedly due to his being a close friend of Nancy Reagan. Dr. Graham was tall, lean, and very academic in appearance. I could easily picture him in corduroy pants, wool jacket, bow tie, and vest lecturing to an economics class at one of the prestigious Ivy League schools. He seemed to be so misplaced yet was savvy enough to step in and lead the space agency until a permanent administrator was appointed. During one of the several STS-61C launch countdowns, I was asked to tour him around the Launch Control Center late one evening. He asked very intelligent questions, was enthused to see the complex launch processing hardware up close, and displayed the utmost courtesy in shaking hands and exchanging pleasant remarks with employees. Being far removed from the NASA Headquarters operations at that stage of my career, I would have been more than pleased to see Bill Graham appointed as NASA's Administrator.

Jesse Moore was a gentleman's gentleman. As the Associate Administrator for the Office of Space Flight, he was the responsible person in the NASA Headquarters hierarchy who we felt we "worked for." A great person, Jesse was an electrical engineering graduate with a Master's Degree from the University of South Carolina. As most electrical engineers are, Jesse was sharp! He possessed the added sense of seeming to know who he could rely on and when to pursue an issue in depth. For those two particular reasons, I'm convinced Jesse Moore would have become totally involved with the solid rocket "O" ring problem had someone elevated it to his level. He commanded respect because of his management style. I do not recall one bit of criticism regarding Jesse's style or any decision he made. If he had a fault, it was simply that he was too nice! Probably one of the most maligned men involved in the Challenger investigation, at least the one completely innocent of having made mistakes, was Jesse Moore. Many others received varying degrees of criticism, accusation, and dishonor. Some were practically "rode out of town on a rail;" Jesse Moore deserved to be elevated to a position of more leadership, for he was a regular guy and a dedicated government service employee. Most of us at my level were surprised to learn after the Challenger accident that Jesse had been named as the new Center Director of the Johnson Space Center on January 23, 1986. After the first weeks of the Challenger investigation, Jesse spent a few short weeks at Johnson, but never actually took over as its Director.

Another extremely popular and well-respected participant in the Challenger activities was the KSC Center Director Richard (Dick) Smith. An Auburn War Eagle, Dick had distinguished himself at NASA Headquarters and earned the respect of every top manager in the Agency. Dick has gained a lot of renown for being the man in charge of assuring the Skylab module did not endanger a population somewhere on earth after its fiery reentry into the earth's atmosphere. Always cheerful and encouraging, Dick was a supreme delegater. Mr. Smith once told me after one of the smooth launches in late 1985, "Gene, I think you're going to be the coolest Launch Director we have had so far." Words like those have immeasurable value to an employee. Dick pictured his role as Center Director to be one to set policy, rule in disputes, and put the Center's best face forward to NASA, the public, the media and academia. He was an ideal Chief Executive Officer. A fellow Christian, we shared a kindred spirit of appreciation for each other.

Dick Smith's counterpart at the Marshall Space Flight Center was Dr. William (Bill) Lucas. My association with Dr. Lucas was never anything but cordial and professional. I considered him to be the ultimate leader for a large government research and development facility like Marshall. Dr. Lucas was a strong conservative Christian layman, a leader in his church in Huntsville. He was a family man and was known for his values and ethical thinking by the Center's employees. Due to the discoveries of the history of the SRB O-ring anomalies, the Marshall Center took a real beating from the press and by the numerous investigative groups named to look into the Challenger accident. Bill Lucas took a lot of the beating. I think a lot of it was undeserved. During the investigation phase of the Challenger accident, an employee at Marshall wrote and distributed what was known as the "Apocalypse letter." This short memo totally denounced the top level management at Marshall. It singled out Dr. Lucas as a tyrant who accepted no idea but his own. It implied in so many ways that Dr. Lucas' leadership at Marshall had led to the type of management that allowed the SRB "O" ring problems to be hidden. The letter was widely circulated around NASA and the media. I never learned who authored this libelous epistle, but the damage it brought to Dr. Lucas' career was devastating. Suddenly, "enlightened people" came forward to tell what a strong, non-yielding bureaucracy had always existed at Marshall under Dr. Lucas. "You didn't question Dr. Lucas and his commanders at Marshall. You didn't air your dirty laundry to the media. He ruled with an iron fist. You didn't make any decision or tell anybody your problems until first taking them up the long chain of command until they got to the ninth floor, Dr. Lucas." A lot of this was baloney and a lot of the poison letters criticizing Marshall management and Dr. Lucas were "sour grapes" employees who have never worked a day without sharply criticizing someone or something that they didn't like. Marshall certainly had its share of organizational and managerial shortcomings, but this exists in all private companies, government agencies, or public institutions of such size. The foibles that allowed Challenger to happen were the result of many failures on the part of many responsible people. Bill Lucas took more of the blame for the accident than he justly deserved!

The Marshall Spaceflight Center had a strong Shuttle projects team under Stan Reinartz. Jovial Billy Taylor was my good friend and Main Engine Project Manager. Porter Bridwell, later to serve as Marshall Center Director, was project manager of the External Tank and a real veteran. The one person who took the biggest blame for Challenger, often slandered by elements of the media, was the Solid Rocket Booster Project Manager, Larry Mulloy. A slender look-alike of Walter Matthau, Larry was a unique project leader with innovative ideas and a strong business sense. On January 14, only a week or so before the Challenger accident, Bob Sieck and I met with Larry to start planning an innovative and unheard of proposal: to actually transfer some of the Marshall work and authority to the Kennedy Space Center in order to cut costs. We were beginning the process of transferring the booster recovery and disassembly work functions under Kennedy's control. This would be an additional work responsibility for my launch and landing organization. We were firmly convinced that any work not requiring a strong designer involvement should be transferred from the design center, Marshall, to the field center, Kennedy. Larry was one of the first project managers to see the cost benefits of this type of operation. He saw it as an opportunity to free more project money for other significant SRB needs. To the world outside the aerospace industry, those unfamiliar to the inner working of the space program, there is no real comprehension of the complexity and enormous responsibility of the people who hold such critical positions. Such a position was that held by the ET, SRB, and Main Engine project managers. The Orbiter project manager Dick Colonna had a tremendously demanding job. The Launch Director and the Director of Shuttle Operations' responsibilities, though willingly and proudly assumed, offered nothing but stress and more stress. Larry Mulloy must have reflected back over his government career and often wondered if the stress and risk he assumed were ever worth the rewards. Positions in America's top corporations at this level, holding this much responsibility, bring salaries two or three times greater than those of government executives. Unfortunately, in positions of this nature, good honest, ethical men and women are called upon daily to make judgment decisions which, if wrong, can lead to destruction, failure, or budget chaos. In my humble opinion, such was the fate of Larry Mulloy. Larry looked at the facts, heard the objections, aired his questions, and voiced his complaints. He made a very bad judgment call. Everyday in the space program, managers make decisions that are questionable. There are always people who do not agree with managers' calls. These people need to be heard. Any factual data must be analyzed and dispositioned. But good managers cannot afford to let every person who has a "bad feeling" about a problem have a vote. Just because someone has a notion that something isn't exactly right doesn't give them license to stop the wheels of progress. It appears Mulloy had a wealth of negative data, a lot of people telling him not to fly, and did not have the positive margins needed to give the "O" rings a clean bill of health. In retrospect, all accidents and disasters seem to call forth multitudes of individuals and groups just waiting to criticize the decision makers. Mulloy made a poor call, but we should recognize the pressure to which we so often subject our leaders.

Harry Truman was the last voice in deciding to use the A-bomb against Imperial Japan. He was put in a risk position, subject to extreme scrutiny. He weighed the situation, the facts, and the advice of his aides and made the call. As long as mankind exists on earth and scholars are free to write, Harry Truman will be loved for his excellent decision or hated for his dreadful decision.

Larry Mulloy did not make a decision to jeopardize the lives of seven great Americans. He made a judgment call on a technical situation that proved to be wrong. It proved to be fatal!

NASA by the nature of its mission must make very difficult judgment calls which are always subject to outsiders' critique and criticism. A world of Monday morning quarterbacks are always looking across the line of scrimmage at NASA in every endeavor the agency pursues. If NASA managers make a determination to move in one direction and fail, they are asked why they made such a dumb move. If you select the complete opposite and fail, you are asked the same question, "Why did you select that course of action? It doesn't sound reasonable." And never is the praise given for success as great as the criticism received from failure!

The Center Director position at the Johnson Space Center was vacant at the time of the Challenger accident. Even so, the Johnson Shuttle role was significant as three strong managers were in important positions: Arnie Aldrich, Manager of the National Space Transportation Program Office, Cliff Charlesworth, Director of Space Operations, and Aaron Cohen, Director of Research and Engineering. In later years, I became closely associated with Aaron Cohen, one of the very nicest people I have ever known!

With all apologies to my good friend, Arnie Aldrich, he would have been called a nerd today on a college campus. With thick glasses and a scholarly look, Arnie was a highly intelligent strategist who appeared to know everything and be on top of it at all times. As the Chief Project Manager on the Shuttle, he knew the details of the Orbiter's electronic brains as well as anyone. This gave him a heads-up knowledge of the core element of the Shuttle. A graduate of Northeastern University in Boston, Arnie rose through the NASA ranks from one highly responsible job to the next. As I learned to do consistently in my job as Director of Launch and Landing Operations, I relied on Arnie to give the final word on the condition of the Shuttle for launch.

The Deputy Director at the KSC was the "Dancing Bear," Tom Utsman. A brilliant engineer, Tom was a mentor and role model to a lot of managers who climbed the ladder of success at the Center. A big bull of a man, Tom had started his professional career as a petroleum engineer. A Purdue Boilermaker, he spent a lot of his early engineering days on the oil rigs in the Gulf of Mexico. Coming to NASA as a design engineer, Tom helped the Von Braun and Kurt Debus team design and build the huge Apollo launch facilities. Tom progressed to become director of the Shuttle Operations Directorate which he handed off to Bob Sieck when he became Deputy Center Director. Director Smith relied on Utsman to keep his hand on the Shuttle pulse and be his advisor on Shuttle matters. Tom spent many hours checking and double-checking our planning and strategizing every detail of what challenges were in front of us. It was a real asset to have Tom Utsman and Bob Sieck ahead of me, advising me while still allowing my organization to carry out its mission. In January 1990, it was my privilege to succeed Tom Utsman as Deputy Director of the Kennedy Space Center.

Bob Sieck, the ultimate Launch Director, had been a friend and confidante since the days of the Gemini program when we were biomedical instrumentation engineers in the same small organization. I assumed the job of Shuttle Project Engineer from Bob when he became the first Shuttle Flow Director and later assumed the Launch Director position when Bob became the Director of Shuttle Operations. Our careers have always interlaced. We were the two engineering managers sent to Edwards Air Force Base in 1976 to conduct the Approach and Landing Tests on the first flyable Orbiter, the Enterprise. Bob was a top graduate of his electrical engineering class from the University of Virginia, the son of a career federal government employee, raised in Falls Church, Virginia. Brilliant to the point of being unbelievable, Bob Sieck could grasp a technical situation in depth faster than any individual I've met. He understood how science and machines work and this talent, in my opinion, is the optimum criteria for a good Launch Director. An avid sports car enthusiast, Bob liked to race his Formula V sports car around the circuit in the deep South when his job and busy schedule permitted. He says it let him relieve the stress of the space program. Bob flourished in the leadership role as Launch Director and with the media at press conferences. He detested the day-to-day mundane role of a manager to which I had become accustomed. When the Challenger exploded, I know that all those who have served in the role of a Launch Director must have felt the pain and loss that I felt at that moment. I know Bob Sieck did!

A pyramid management structure characterized how the NASA hierarchy and organizational controls were established for the Shuttle Program. Jesse Moore was referred to as Level I, the top level of NASA management for Space Flight, a NASA Headquarters Washington function. Arnie Aldrich was the overall Shuttle Space Transportation System, or STS, Program Director, Level II. Arnie was actually the day-by-day boss of the entire Shuttle program. Only in the case of Flight Readiness Reviews and when the last vote to launch was given, did the established protocol pass the final authority to Jesse Moore of Headquarters. So for the Challenger launch process, KSC actually received Flight Readiness go-ahead and launch commit direction from Level I, Mr. Moore. The other level and lowest on the organizational chain was Level III, the projects elements such as the Solid Rocket Boosters under Larry Mulloy and the Shuttle Operations under Bob Sieck. Theoretically and ideally, all and every Shuttle problem would flow up this chain of command as high as its importance should dictate. This was not a bad span of control arrangement; it had been similarly used successfully on the Gemini and Apollo programs. I sincerely believe the system was only faulty in that it depended too much on the judgment of a lot of key people at the three management levels. If someone along the ladder made a bad judgment, the result could be nothing short of catastrophic!

My good friend Jim Harrington played a significant role in the Challenger saga. Jim was Challenger's Flow Director, the one person we relied on to follow the Orbiter all over the world as it was processed and recovered. He had no direct reporting employees, but in a sense, everyone at KSC working on the Challenger reported to Jim. After Challenger was mated to its SRBs and External Tank, Jim was Flow Director for the entire Shuttle stack. We gave our Flow Directors a lot of responsibility. Jim, a Florida native, was a graduate of the University of Miami and one of our best operations engineers. After Challenger and after assuming jobs of increasing responsibility, Jim was appointed Shuttle Launch Director, a job he was obviously qualified and prepared for. Highly visible and respected, Jim carried out the Launch Director duties in a professional manner.

I often remarked to friends that Jim could easily play himself in a space movie. He actually did so in a television series, *The Cape*, about the Shuttle program. I have a penchant for imagining who would play the role in movies of true life people I have known. I would certainly see how Gary Sinese could don a pair of glasses and look exactly like the seasoned Launch Director Jim Harrington. Jim was another long-time comrade who shared with me the pain and anguish of the Challenger experience.

No single person has contributed more technical support to the Shuttle program than super engineer Horace Lamberth. A native of Tennessee, Horace is one of the excellent athletes at the Kennedy Space Center who could have attained a successful career as a professional baseball player. The KSC launch team always looked for Lamberth's opinion and approval before any technical problem was fully resolved. Horace's approval was an unwritten requirement that must be part of any closure rationale on a Shuttle ground problem report of any significance. Horace's sharp analytical mind, his knowledge of the Shuttle technology, and his dedication to safety were all unbeatable characteristics of what NASA wanted its top Shuttle engineering chief to possess. As Launch Director, I never seriously considered launching without the assurance that Horace Lamberth concurred. I have not crossed paths with many people during my career who I consider close to being indispensable. Horace Lamberth is by far the closest. We worried a lot at NASA about the loss of corporate memory. This is the situation where the brains of an organization are slowly drained by retirement, by resignations, or by transfer of individuals to higher level jobs. Horace was always content to act in the chief engineer's role despite numerous opportunities to assume executive positions. NASA and the contractors continued to value this man's abilities and kept him in the critical path to review all major Shuttle engineering problems. Replacing a person of Lamberth's capability should be a goal outlined in a contractor's strategic survival plan. A less notable quality of his amazing engineering mind was his obvious addiction to a simple soda concoction called Dr. Pepper.

I could go on for volumes describing the stars and superstars of the KSC launch structure and the NASA management team at the time of Challenger. I will only speak of two more players.

Norm Carlson, a country boy from Oklahoma, an Oklahoma State University graduate and a patriotic, flag-waving space fanatic was the best test director to ever don a headset. The Launch Director sits in his lofty chair and directs the thousands who support a launch. But the "nitty gritty" nuts and bolts are left to the NASA Test Director. We often joked about what we would expect if a grenade were tossed into the test conductor row in the firing room.

The NASA Test Director would be expected to throw his body on it and take the damage. Norm Carlson was the steadying influence in a launch control room environment. He exuded an air of confidence and control. He assumed the role of father, priest, and gifted mentor to hundreds of aspiring engineers looking to learn the ropes. I often wished I had learned my firing room lessons under Norm's guidance. When NASA speaks of retaining the corporate knowledge of its more experienced employees, one of the first examples mentioned is Norm Carlson. Norm had a talent of simply looking in a special way at a younger test director and getting instant attention or change in behavior. In a lighter sense, some say Norm's greatest contribution to Shuttle launches are his beans and cornbread. Early in the Shuttle days, Norm concocted a scrumptious recipe for a pot of beans flavored with ham slices and onions. He brought a couple of pots to an early launch as a post-launch celebration snack. They hit the favorite foods list like McDonald's hamburgers a few decades ago. The piercing aroma of these beans usually began to seep into the cracks and crevices of the Launch Control Center about six hours before expected T-0 time. The number of pots prepared grew and grew until now the launch processing contractor funds the dozens of pots of beans and dozens of trays of cornbread. For those never having eaten beans and cornbread at 1 a.m. or 3:30 a.m. in the morning, you are too sheltered. Some of the older folks in the Shuttle program have said, "When you are celebrating a successful Shuttle launch, the food you're eating doesn't matter at all. Anything would taste good in that situation."

I recall the words once spoken by Frank Caldeiro, now an astronaut, "The best reward I can receive for my work at KSC is the taste of beans in my mouth when I drive home after a successful Shuttle launch." Thanks to the dedicated years of service by Norm Carlson, we have a corps of great test directors on the team. Thanks to Norm, we have a legacy of excellence and pride in how we conduct launch countdowns. But as many would agree, thanks to Norm, we have a lasting tradition and the recipe for the very best beans and cornbread in the world.

The last person I would mention as a participant in the Challenger story as it evolved at the KSC was the Challenger Project Engineer,

Pandora Puckett, affectionately called "P Square" by friends. A long-haired willowy blonde, Pandora was one of my favorite people. We were on the same wave length. We both came from broken homes and struggled to attain our college degrees. Another graduate of the University of Miami, P2 had a knack for priorities. She always knew where to place emphasis and concern and was an excellent project engineer. I was very proud to have been able to select her as the first woman Orbiter Project Engineer when I was the Project Engineering Office Chief. I have always been a strong supporter of women and in some positions I would rather have one woman than a half dozen men. Pandora put her heart and soul into Challenger and often remarked to me, "Challenger is the best Orbiter and the Challenger team is the best Orbiter team." In my position, I had to avoid showing preference, but her boasting was hard to argue with. The day after the Challenger accident, Pandora came to my office. We hugged each other, and just sat in silence for ten minutes or so with tears in our eyes. Neither spoke a word. It was as if we were having our own special service in quiet honor of the fallen crew. Like all these people I have tried to personalize in words, Pandora became a close friend forever. I was so happy when Pandora announced her engagement to another super comrade Bob Crippen. "Crip" and Pandora later returned to KSC as the Center Director and First Lady and I had the privilege to serve as Deputy Center Director for him over a three year time period.

I think God provides friends for us as we pass through the stages of our lives. Some friends are workmates, fellow laborers in various activities. Some friends are those we respect, depend on, and would sacrifice for if they needed anything. And then some are special friends. These are those I am sure God intended to touch our lives or be touched by us. Not every acquaintance is a special friend, for God did not intend it that way. I think we have in part done the will of God when we recognize and love those special friends of our life's journey. One of the great joys that I envision to be part of eternity in heaven is to live forever with Christian friends we have come to love in this short, short journey on Earth.

CHAPTER 16

Seven Exceptional Americans

Many times when reliving my experiences and those associated with the Challenger accident, I have considered the idea of whether a recollection of the crew's activities should be the subject of which to write. Do I as an associated player in the gigantic operation called Space Shuttle have a right to give accounts or my impressions of these fallen heroes? The same answer keeps coming back over and over! Yes, these were real people working with other real people like me who were dedicated to the accomplishment of a great technological endeavor. The crew was special because they were the most visible. They were the envy of all of us in the program, yet we all recognized our individual contributions. All Shuttle team members are proud of their involvement, dedicated to making the program successful, and hold the welfare of the Shuttle crews as their first priority. Like all successful endeavors with challenges and risks, someone has to take the responsible leadership roles. It was my destiny to be given the honor of being one of those people. We often downplay our contributions, either because we want to avoid sounding pompous or we fear we may overstate our actual involvement. I have always been prone to understate my own contributions, but I hope this accounting is as close to the actual historical facts as possible.

In knowing these crew members to be special since they were chosen to fly, I feel that I would not impose upon them or their families in giving personal accounts of my own association with them. We all had a job to do; Challenger greatly affected our lives; most Shuttle personnel close to the Challenger were touched significantly. Their careers were altered. Some of us lived to tell the stories that few people were able to share. There are certainly discreet events in the history of our country which deserve recording for historical purposes. Challenger was one such event in the 35 years of space exploration.

As I assembled the chapters of this book, I purposely chose January 28, 1997, eleven years after the accident, to record my remembrances of each of the crew. At the point of this writing (January, 2006), it has been twenty years since the event. Because they were all special in their own way and yet so different, it would take chapters to describe each member of the crew, so this account is brief.

The Challenger crew that perished on January 28, 1986 was the most diverse crew NASA has ever flown into space. Being a strong advocate of diversity and equal rights, I was filled with special pride when the STS-51L crew was announced. Two crackerjack white male test pilots, a brilliant young Jewish woman, a black physicist with strong technical credentials, a company engineer specializing in science, a great amicable flight engineer of Asian Pacific descent, and, of course, an All-American lady, a science teacher of extraordinary grace and skill.

I am persuaded that our righteous Almighty God does not judge us by the color of our skin, by the origin of our birth, or by whether we are male or female. For that basic reason, I considered the diversity of the Challenger crew to represent all that was good about America, all the progress we had made over the two centuries of our existence to assure equality for all mankind.

I first met handsome smiling Dick Scobee during the early testing of Columbia for the STS-1 flight late in 1980. We were a few days before taking the Orbiter to the vertical position to be attached to the external tank and solid rockets in the Vehicle Assembly Building. As the Orbiter Project Engineer for Columbia, I had a lot to say about what tests we ran on the Orbiter, where and how we ran them, and the myriads of procedures and equipment we used in our test program. The ultra-conservative John Young was commanding the first flight of this technological marvel. He wanted us to perform tests of the Orbiter's flight control surfaces with the Orbiter in a vertical orientation. He called on a fairly new astronaut at the time, Dick Scobee, to come to the Cape to make certain that we were doing not only the vertical testing that he wanted, but that we also had operational television cameras strategically located to record every move, flutter, or whatever motion the aero surfaces might undergo. Dick had called me concerning John's wishes and the two of us agreed to walk down the VAB and determine where the TV cameras should be located. I was impressed by Dick's sincerity and down-to-earth attitude from the first moment we shook hands. It was soon obvious that everyone who ever met Dick Scobee would like him and would become a friend.

We trudged up and down steel stairways, climbed all over the massive beams in the VAB trying to get a bird's eye view of the numerous control surfaces through the openings in the structures. We spent at least three hours doing a survey and making notes as we progressed. By morning's end, we had a good perspective of where to tell our TV personnel to locate the special cameras. Most meaningful to me as I look back on the morning we spent together was that we gained a mutual respect for each other and started what was to be a lasting friendship. I sincerely feel it was God's will that Dick Scobee and I met long before we were so eternally bonded by the Challenger accident.

Francis R. (Dick) Scobee was a native of the state of Washington, born in 1939. After graduating from high school at Auburn, Washington, Dick enlisted in the U. S. Air Force and was trained as an aircraft engine mechanic. A top performer and achiever, Scobee attended night classes to obtain college credit. In those days, the military term for an enlisted man who worked for his college degree while in service was "mustang." A mustang is considered a proud steed in modern times, strong, fast, and reliable. Dick was a mustang in the truest sense of the word. Dick's strong determination resulted in a degree from the University of Arizona. From there he was accepted for flight training by the Air Force. Tall and thin, Dick Scobee reminded me of a hero of my youth, John Wayne. Dick's strength of exuding confidence was always quite evident. I was so pleased when NASA managers at the Johnson Space Center announced Dick's first assignment as commander for Challenger's mission STS-51L.

One Friday while having lunch with Dick in the cafeteria of the Launch Control Center, he was handed a note by one of the crew support team members. He ran to the nearest phone to call home to Houston to find that his son Rich had been injured in a motorcycle accident. The accident, thankfully, was not too serious. Being a dad myself, I knew from the expression on Dick's face and the concern in his eyes that he was a dedicated father who loved his children. After hurriedly finishing his lunch, Dick said goodbye and rushed off to get a T-38 aircraft ready for an early return to Houston. This same son was later to fly the missing man jet maneuver over the Super Bowl in January of 1996 to commemorate the tenth anniversary of the Challenger accident.

I honestly do not recall how I met Mike Smith for the first time. My impression of Mike was that he was clean-cut, all-American, bright-eyed and super smart. We also became good friends. I hesitate to use the term "close friends" with any of the Challenger crew members, because we never socialized, never went fishing, or to a ballgame together. When we ate together, it was usually part of an official function, a planned event, or during lunch breaks after a long morning of countdown procedure review. Out of our associations with the flight crews grew a mutual respect and admiration for the jobs we each held and our individual responsibilities in the space program. Were I to imagine what Mike Smith was as a young boy, I would think of Tom Sawyer: ingenious, fun-loving, and full of life. When I was to celebrate my fiftieth birthday in September 1984, one of my fellow project engineers invited Mike to drop in if he were in Florida at the time. The party was a "bring a dish" lunchtime gathering and I was pleasantly surprised to see Mike show up. I later learned that he had found a special reason to be at the KSC that day to honor a friend turning the corner of middle age. Mike always brought a glow of sunshine when he took part in any event. I learned to respect his astronaut skills as well as his winning ways. I look upon the memory of Mike and would use a well-worn cliché: "He was a man that any father would be proud to have as a son."

A star football player in high school, Mike was very close to his older brother Pat. The two brothers were starters on the Beaufort High School football team in the fall of 1962, the year I graduated from college and joined NASA at Cape Canaveral. When Mike graduated from Beaufort High School in 1963, he was an all-star quarterback on the football team. He went on to the U.S. Naval Academy, graduating in 1967. A decorated Vietnam War hero, Mike flew hundreds of aircraft carrier missions, winning more than twenty commendations. Mike married his lovely wife, Jane, an airline stewardess he had met as a teenager in South Carolina.

I had known Ellison Onizuka quite a few years longer than the rest of the Challenger crew. I first met him during a two year tour of duty with NASA at Edwards Air Force Base in California. We were having a lot of success doing flight qualification testing of the first Orbiter Enterprise in 1976-77 at Edwards. Ellison was a young Air Force Captain stationed at Edwards participating in flight testing of several high performance jet aircraft as a flight test engineer. I began to notice two young captains sitting in the back of the room and listening intently each day as we held our daily 2 p.m. operations and scheduling meeting. Dressed in Air Force khaki uniforms, one of the two sharp officers was an Asian Pacific Islander. Out of pure interest and a desire to fly, Ellison and his friend attended our meetings just to learn about what we were doing and how we were progressing. For several weeks, I assumed the two handsome well-groomed officers were there to critique us or to support any special requests we might have. Upon arriving a little early for the meeting one day, I walked back to the two men, extended my hand, and introduced myself. At once the trademark smile that Ellison was so well-known for and the bubbling personality made me feel like an old friend. The two men asked a lot of very tough and challenging questions and I soon discovered that they knew a great deal about the Orbiter and its mission. From that day, we conversed a lot and became friends who only saw each other under those conditions. I remember vividly the day Ellison said, "Someday I'm going to fly into space in one of NASA's shuttles. I'm going to be an astronaut." Just a short ten years later, my good friend who I had grown closer to since that first conversation at Edwards was to be a member of the

Challenger crew. I was to later learn that Dick and June Scobee were serving an Air Force tour of duty at Edwards at the same time. Over the years, I am continually amazed at how the lives of people will cross in some insignificant way and then become a very significant remembrance several years later. Such was my great relationship with Ellison Onizuka!

I did not know Judith Arlene Resnick well. Her close friends affectionately called her "J.R." I had met her briefly when she flew her first Shuttle mission aboard Discovery Mission 41-D in 1984. A gifted pianist, she loved math and science and excelled as a student at Firestone High School in Akron, Ohio. This beautiful young woman was brilliant, having earned a doctorate in electrical engineering from Carnegie Mellon and the University of Maryland. Resnick worked as a scientist for RCA and Xerox before joining NASA. She was an expert in operating the Orbiter's manipulator arm and precisely deploying or retrieving deployable payloads from the Orbiter cargo bay. I met Judy again in some debriefings after the first Main Engine abort on the pad when we were trying to launch Discovery on its first mission in June 1984. I remember how she maneuvered her lithe, thin frame through the Orbiter's crew hatch almost in a dead run when we requested the crew to perform a hasty egress. Flying with veteran commander Hank Hartsfield on her first mission, Judy was quick to point out to us how exciting it was to experience a main engine abort. She also related how exiting the hatch so quickly was easy if you were properly motivated. One of the first women chosen for the Astronaut Corps, Dr. Resnick was a remarkable individual, exemplifying women's contributions to our space efforts. I often heard discussions, none officially substantiated, regarding Dr. Resnick's assignment on Mission STS-51L. Many space workers were of the opinion that the crew assignment process had selected Resnick to fly along with the school teacher as a stabilizing, experienced influence to make Christa feel more comfortable. Like all hall talk, I ignored it as being credible unless I had been personally involved in the decision. Thank God, I was never involved in selecting the five or six astronauts to fly a Shuttle mission.

The two crewmen aboard Challenger that I wish I had gotten to know better were Greg and Ron: Gregory Bruce Jarvis and Ronald

Erwin McNair. I had met Ron, again through pre-launch briefings we presented to the crew in early 1984 as he was a member of the crew who flew on Challenger's maiden flight. Ron picked cotton and tobacco as a youth in Lake City, South Carolina. He played four years of varsity football in high school, two years as captain. Ron was only the second African American to fly into space. A sharp physicist, Ron was a magna cum laude graduate of North Carolina Agricultural and Technical State University and prestigious MIT. A deeply religious individual, Ron was a deacon in his church in Houston where he met and married Cheryl Moore. As I learn more about life and maturity, I come to realize the thousands of missed opportunities we have to share our Christian faith with others. I would have loved to spend a few golden moments getting to know this man better.

In my own opinion, Greg Jarvis was as "innocent" a victim of the Challenger accident as was Christa McAuliffe. Greg was an exceptional honors student at Mohawk High in New York State. After graduating as an electrical engineer from the State University of New York, Jarvis rose to the rank of captain in the Air Force before working as an engineer for Raytheon. A practical engineer, flying for his company (Hughes) to operate fluid systems experiments in space, Greg had reportedly been "bumped" from the Columbia STS-61C flight to accommodate a ride into space for Congressman Bill Nelson. Some sources maintain that Greg was not bumped but rescheduled because the hardware he was to operate was not ready to meet the earlier flight date. I remember distinctly the evening that we had barbecue with the crew and after greeting all the guests, Greg found a quiet corner of the room where he could talk technical details of his upcoming flight with one of the experiment's designers. Greg was an intense, dedicated individual. I heard him tell this visitor how they would get together and review/debrief the operations as soon as Greg returned and could get back to their laboratory in California.

I suppose I would describe Sharon Christa McAuliffe best as a "wide-eyed innocent." No amount of coaching or brain-washing could have ever impressed upon this New England lady, mother of two, and loving wife of a successful lawyer, the intense danger that could evolve from a Shuttle mission. She never proposed herself as an astronaut, rather as a pioneer who was excited to be the first Teacher in Space. She was poised to reawaken America's students to math, science, and engineering by teaching science from space. A product of low income housing near the heart of Boston, Christa attended Marian High School, played high school basketball, and sang in the school's production of *The Sound of Music*. Christa's quest for learning led her to receive her master's degree in education. She accepted the selection as Teacher in Space with humility, having won the job over 10,000 other applicants. At the time of selection, she was teaching history and social sciences at a Concord, New Hampshire high school.

When President Reagan announced the Teacher in Space program in August 1984, he said, "When the Shuttle takes off, all America will be reminded of the crucial role teachers and education play in the life of our nation. I can't think of a better lesson for our children and our country."

The three occasions that I was associated with Christa were not such that I could talk with her at length. Our traditional barbecue at the beach house was a chance to see her with her mom and dad and to observe how she was so close to them. I do not recall her husband Steve attending. I do recall the glow of anticipation and wonder that was expressed on her face. I suppose my most impressive remembrance of Christa McAuliffe was her sincere, radiant smile. Often I found in my long career with NASA, we cross paths in life with great people who we later wish with deep conviction that we could have gotten to know better. Such a person was Christa McAuliffe. I have always believed there are classes of occupations in America that we cherish more than others. I'm concerned that the list gets smaller with time. We recognize certain positions as special: doctors, ministers, and school teachers come to mind. In my younger days, politicians, law enforcement, and lawyers were in this group. I consider Christa McAuliffe's sacrifice to have been more difficult to accept due to the special nature of her role in this flight. I would have certainly felt the loss of any spacecraft crew member, as I had earlier felt the pain of the Apollo 1 fire in 1967. But to lose an American civilian, the first real "passenger on a space-going airliner" was even more painful.

As events bring the memory of the Challenger accident to mind on frequent occasions, I realize that we have all lost someone who was part of each of us, someone with whom we could closely identify. I want to remember this tremendous lady as she appeared on the February 10, 1986 cover of *People* magazine: bright-eyed, radiant, smiling with anticipation. Christa held the torch high for education when she said, "I touch the future, I teach."

I thought seriously of visiting the hometowns of each of these seven fallen heroes and conducting an in-depth personal study of each of their childhoods and early experiences. I certainly could have researched thoroughly and given keen insight into who they were, how they grew up, and what they did before they joined NASA. But I decided that I should present the crew as I knew them. Our lives only touched briefly. Someday in glory we shall join hands and hearts again. I do not understand why Challenger had to happen or why God permitted it to be one of the dark chapters in America's story!

CHAPTER 17

The Challenger Cargo

After a lot of press and publicity concerning the payloads on Challenger, very little was written about them following the fateful mission. In retrospect, the STS-51L payloads were quite significant and impressive. Largest of these payloads was the Tracking, Data Relay Satellite, TDRS-B, a sophisticated communications satellite which was to be another in a series of NASA communications stations designed to provide near 100% tracking, data and voice communications between the Orbiter and two or three ground stations. This Hughes-built satellite was attached to an Inertial Upper Stage, IUS, a two-stage rocket motor used to boost the TDRS communications satellite into a geo-synchronous orbit, approximately 22,300 statute miles above the earth. TDRS-B was to be the second in a series of several similar satellites to be strategically positioned around the earth. TDRS-A, the first in the series, had been placed in orbit on an earlier mission flown by Challenger, its maiden flight in April 1983. This first TDRS was located over the Atlantic Ocean at 41 degrees west longitude. The second TDRS was to be positioned over the Pacific Ocean at 171 degrees west longitude.

One of the "best performances by an acronym originator," SPARTAN, was the other payload of some size and weight in Challenger's payload bay. The scientific objective of Spartan-Halley was to measure the ultraviolet spectrum of the Comet Halley as the comet approached the point of its orbit that would be closest to the sun. This free-flying instrument, intended to make ultraviolet measurements and photographs of Halley's Comet, was named SPARTAN/Halley, Shuttle Pointed Autonomous Research Tool for Astronomy/Halley's Comet.

A very intelligent civilian scientist, Greg Jarvis, an employee of Hughes Aircraft company, was aboard Challenger to conduct six fluids experiments. They included fluid position and ullage, fluid motion due to spin, fluid self-inertia, fluid motion due to payload deployment, energy dissipation due to fluid motion and fluid transfer.

Designated by some as teacher-observer but officially flying as a Teacher-Payload Specialist, Christa McAuliffe was to perform several space experiments for high school students. A good portion of this was to be covered live on television from Challenger to the ground. The Teacher in Space experiments were the effects of gravity on hydroponics, magnetism, Newton's laws, effervescence, chromatography, and the operation of simple machines. Christa was also trained to assist with three other onboard student experiments; the study of chicken embryo development in space, research on how microgravity affects a titanium alloy and a crystal growth experiment.

One of these student experiments, Chicken Embryo Development in Space, was a part of NASA's Shuttle Student Involvement Program which enabled bright students to suggest, design, and fly experiments that were appropriate to be conducted in a microgravity environment. John Vellinger, a sophomore mechanical engineering major at Purdue when Challenger flew, had first proposed his "CHIX IN SPACE" experiment to NASA when he was a ninth grade student in Lafayette, Indiana. CHIX IN SPACE was designed to study the effect of weightlessness on twelve developing White Leghorn chicken eggs.

This particular experiment held personal significance for me at the time. The sponsor of the embryo experiment was Kentucky Fried Chicken who had secured the help of Goldkist, a Georgia-based farm and poultry product company renowned across the southland for its quality. Harold Andrews, a Vice President for Goldkist had sent me a very formal letter, which in a light vein asked me as Launch Director for Shuttle to be real careful with the eggs he was providing to fly on Challenger's 51-L mission. The uniqueness of this request was that in October 1985 Harold Andrews' fine son, Steve, had called the Launch Director's office at the Kennedy Space Center and made another official request. Steve had asked my blessing to propose to my beautiful daughter Karen. He had subsequently given Karen a lovely engagement ring, and they were to be married in July 1986. I had responded to Harold's letter assuring him that we would certainly take extremely good care of his eggs on Challenger's flight. It was probably the first and only time, in a sense, that the Launch Director had a somewhat vested interest in the successful mission of part of the Orbiter's cargo.

CHAPTER 18

The Days before Challenger

There has always been a fairly disciplined set of guidelines for determining if one of America's manned launches is ready to launch. Such was the case for the Shuttle launches prior to the Challenger mishap. One of the major weeklies, *Newsweek*, called this a painstaking process. The Challenger had landed at the DFRF in California from its October 30, 1985 launch and went through the normal deservicing, ferry, test and reconfiguration to fly the next mission STS-51L. On Monday, January 13, 1986 we ran a nominal and successful Countdown Demonstration Test with the Challenger seven, the very next day after launching Columbia on its STS-61C mission. On Tuesday, January 14, 1986 we held the Launch Readiness Review chaired by our Center Director, Mr. Dick Smith. I had sent a memorandum from my office, under my signature, defining January 15, 1986 as the date we would hold the STS-51L Flight Readiness Review at the KSC. We were conducting business at a very busy schedule pace. It appeared that every day was filled with a launch, a major test, or a critical review. For several reasons, the Challenger STS-51L launch date was slipped from late December 1985 to January 23, 1986. The main reason for slipping the launch dates was the "interference" caused by Columbia's STS-61C launch problems. The weather and hardware scrubs experienced in the December 1985 to January 1986 period on Columbia played havoc with Challenger's schedule. Our KSC workload forced us to slip until January 26, 1986. We held our normal Launch minus One Day review on January 25, a Saturday. This meeting is established to review any last minute hardware problems or concerns. It also let the launch managers look at the predicted weather for tanking of the External Tanks propellants and the weather predicted at launch time. The Johnson Space Center also reported the weather expected at the Transatlantic Abort Landing (TAL) sites and the CONUS emergency landing sites, those in the continental United States. The weather forecast we heard was horrible! With a forecast of overcast skies, rainstorms, and marginal ceilings, we were advised that the weather would be bad for a Sunday morning launch time around 8:30 a.m. We had experienced many weather worries before Columbia's launch on January 12 and we learned to listen well to our Air Force and civil servant weather advisors. The forecasters were predicting that Sunday would bring terrible weather to the Florida peninsula. After asking a lot of questions regarding this prediction and getting a definite feeling that the weather experts were correct, we elected to forego a Sunday, January 26 launch attempt and shoot for Monday. The weather on Monday was predicted to be clear, cold and balmy as a huge high front was moving in behind the bad weather predicted for Sunday. Weather variability in the northeast region of Florida can be a real problem to work with when you are a Shuttle manager. Rain, wind, lightning, and hurricanes plague you in the summertime. Cold temperatures and winds create problems in winter. I always looked forward to having the October to February time period arrive. During these Florida wintertime days, clear skies are usually present until at least mid-afternoon. I'd always choose an early December launch date if I had a choice.

Following the Challenger accident, a lot of press was directed towards schedule pressure at KSC and that it had been instrumental in causing Challenger's explosion. One source quoted said that we pushed to launch Challenger on January 26 so that Vice President Bush could stop at KSC on Sunday morning enroute to South America. He would be a very high level representative for the launch of the Teacher in Space Program. I was certainly informed that the Vice President might stop by on Sunday. I was never instructed to push in any way to launch. Absolutely no one above or below my level ever even vaguely suggested that we try to launch for Mr. Bush. I never once considered his possible visit to view the launch as having any significance in our launch decision process. Ignoring his possible visit as a factor, we reviewed only the technical and weather considerations. And we scrubbed for Sunday based on inclement weather predicted. The Vice President flew over the south without making a short stopover at the Kennedy Space Center that beautiful Sunday morning.

On Sunday morning I reported to the office as was my practice during the last two or three days prior to a launch when they fell on a weekend. When a countdown is in progress, or holding, you have to cover your responsibilities. There are no days off when a countdown is progressing.

My drive from south Merritt Island to the Kennedy Space Center was like a morning drive in paradise: clear blue skies, temperature slightly below 50 degrees F, and practically no wind. When I arrived at the LCC, the expected members of my staff were there for a countdown on Sunday: those whom I had directed to be there and one or two dedicated souls who are always there when they sense their assistance might be needed. Bob Harris was one such person. Bob was my administrative assistant, or chief-of-staff. He kept our outfit happy and contented while I tried to manage the launch and landing activities. A World War II pilot who had survived a horrible nose-down crash in a small military aircraft, Bob loved flying, those who flew, and those who made things fly. A congenial person by nature, Bob was an invaluable aide to the Launch Director.

Sir, Mr. Dick Young of Public Affairs just called and the press has asked for your comments on today's weather. Dick would like for you to go outside at the time we would have normally launched this morning and comment on what you feel the weather conditions at launch time would have been like."

"Bob, it sure looked good driving in to work. Why don't we go up on the LCC roof about 8:30? Invite Dick Young to come on over and go up there with us!"

At about 8:20 a.m. on January 26, 1986, Bob Harris, Dick Young, and I trekked our way through the maze of offices, doors, stairways, and platforms to the roof of the four story high Launch Control

Center. We stood on the same elevated wooden platform where the close members of the crew families would stand on Monday and Tuesday mornings. We looked out across the five or six miles of scrub vegetation to a clear view of the magnificent Shuttle stack at Pad B. To the southeast was a blue lagoon we called the External Tank turn basin. The water was smooth, not even a ripple. As we scanned the periphery 360 degrees, we observed nothing but clear blue skies and calm. The huge VAB with the American flag and the 200th year Centennial logo on its south walls loomed huge before us to the northwest. The outside temperature was near the mid 40s. There were absolutely no clouds in the blue Florida skies around the KSC this beautiful Lord's Day. What a great day it would have been to launch the Challenger!

"Gene, do you have any words of wisdom to pass along to the media on this weather?" Young asked.

"Well, Dick, what can I say? It would have been a great day to launch, but I guess we missed a good opportunity," I replied.

That's about all the comments I could think of. We had no doubt missed a golden launch opportunity. I had no way of understanding at that moment in time how significant both this missed weather call and the events of the next day, Monday, would be in the fateful saga that was unfolding. I now think often of how instances, events and missed opportunities play such consequential roles in our daily lives.

There is absolutely no way I would ever fault a weatherman's call because I somewhat understand the unpredictable nature of weather itself. So many factors are contributors to or generators of the multiplicity of weather conditions around the globe. This is another of those judgment calls that NASA management has to make repeatedly. Do I proceed under questionable weather predictions? If I proceed and scrub, can I justify the high cost of people and commodities? If I don't proceed and get perfect weather, does it make our operations or the operations of other support elements look inadequate? I never felt bad about having to answer "what if" questions. I supported the premise that to be safe was our first priority, and this allows you to scrub, delay or cancel without unnecessary criticism. I feel "better safe than sorry" is a simple, but strong, tenet to follow in launch-related work. Even though I was disappointed to miss an excellent launch opportunity on Sunday, January 26, I understood as well as anyone else how weather in Florida is quite unpredictable. I marked this one up to experience: simply more data to push back somewhere in your mental data bank. Another lesson was learned, another first for me in my term as Launch Director.

Needless to say, I spent lots of time on the telephone with the weather group that Sunday. I understood how they must have reacted to the totally incorrect weather prediction. I felt total empathy for these folks and actually felt sorry for them. I have always considered respect for others to be a top character trait. I tried to show respect in all associations with others, friend or adversary. There is no justifiable reason to ever disrespect another. I truly respected the weather personnel; they were outstanding, and I told them so.

We had been blessed to have had several outstanding Air Force officers as weather liaison between the Shuttle program and the Patrick Air Force Base Weather Office. Lt. Scott Funk and Lt. Francine Lockwood were two of the best in Air Force Blue. Scott had served as "weatherman" for us in a highly professional manner and subsequently been reassigned to Thule, Greenland as a reward. We found so often in dealing with Air Force officers that their rotation policy often transfers great people away from critical job positions at the most inopportune times. Such was the departure of Lt. Funk. Scott had the dubious honor of advising us on weather during the long string of December 1985 to January 1986 STS-61C Columbia launch attempts. In this particular turn of events, Lt. Lockwood was an exceptionally well-qualified and able replacement. Francine had the misfortune of having to call the weather as "no-go" for Sunday, January 26 and then awake as I did to a "launch-perfect" weather day. Some of her colleagues at the weather station "made her day" by releasing the following informal Shuttle weather announcement:

> *Cape Canaveral, FL (UPI) – Air Force Lt. Scott Funk, the former shuttle weather officer, received good news Sunday at his Thule, Greenland, office when he was told Lt. Francine Lockwood had been assigned to his "command." Lockwood, wearing stylish connectable bracelets, was escorted to a*

> *transport jet Sunday morning at Patrick Air Force Base for the long trip to Greenland just hours after she predicted unacceptable conditions for the Shuttle Challenger's launching. We're calling for good weather for her arrival," Funk said in a telephone interview. "But then again, It could be snowing. It's really hard to say." Funk, whose trademark, "We're cautiously optimistic," endeared him to legions of Shuttle reporters, was transferred to the prestigious Greenland base earlier this month after compiling a record string of predictions that kept the Shuttle Columbia grounded for 25 days.*

As the weather on the morning of January 26 was so great, I joked about what to expect from their prognostications for Monday morning's weather. From all indications, it appeared that the weather on January 27 would be great if we were able to launch in the early part of our window. They predicted that in the late morning hours we might face higher winds than allowable for a return to launch site abort should we have to do so. With a good weather day ahead on Monday, I felt especially good that beautiful Sunday afternoon as I drove home around 2 p.m. Before an early morning Shuttle launch, the launch team comes in to work mid or late second shift the evening before. Another launch support team has activated Orbiter systems, prepped the LOX and LH2 ground systems for tanking, and rotated the huge RSS, Rotatable Service Structure back away from the Shuttle stack. This retraction position protects the RSS and the Payload Changeout Room from fire and blast damage as the fiery Shuttle lifts off the pad. So it was my practice to try to sleep five or six hours the afternoon before launch. I found it reasonably easy to forget the excitement of launch time and slip off to sleep in the coolness of a January afternoon. So I slept until about 9 p.m. on the evening of January 26 and drove in to work about 11 p.m. Everything was progressing well with the countdown, and the prediction of good early morning launch weather was still being reported. I felt extremely confident about our launch probability. The few small nuisance problems, usually somewhere in our aging ground support equipment, only served to keep the launch team members alert and active. We actually welcomed a few non-showstoppers every now and then. This good feeling was not to last long. The day was to become one of the most disappointing of my long engineering career. We actually lost another golden opportunity to launch because of the failure of a simple piece of aluminum. It was an uncomplicated aluminum gadget that any good mechanic in a machine shop could fabricate in a couple of hours. This was the day we scrubbed because of the infamous Orbiter hatch tool!

Another event of much less significance occurred that day. Our Administrative Assistant's wife Marian Harris contributed a helium-inflated balloon to my office as part of our decorations for the Challenger family members. The balloon, like those so often used at birthday parties, anniversaries and similar events, was shaped in the replica of a Shuttle Orbiter. The balloon floated calmly around the office as June Scobee, Jane Smith, and the other family members enjoyed the hospitality and anxiously waited as the launch clock continued to count downward. The balloon was simply another part of the festive atmosphere on January 27. Later it would seem almost as if it had been a forewarning of things to come. The nice outside weather, however, gave us all an assurance that the day would bring a successful launch.

The countdown reached the point where we had ingressed a very happy and ready-to-fly crew. After the communications checks and the crew are securely strapped into their seats, the closeout crew technicians secure the Orbiter side hatch. Then we go through a normal cabin leak and decay check after the hatch is closed. The engineers at the environmental controls console perform this pressure check to confirm the integrity of the crew compartment and the hatch seals against leakage. As the Orbiter climbs out of the earth's atmosphere and the outside pressure approaches zero pounds per square inch, it is imperative that the crew cabin remain at or near 14.7 PSI, or about one atmosphere. A lot of engineers hold their breath through these leak checks because they must be completed within acceptable prescribed limits or the shuttle doesn't fly that day. The astronauts' lives depend on maintaining a cabin pressure of something above 10 PSIA. On the day of January 27, 1986, as the mechanical tech started to remove the hatch tool after these critical tests were successfully completed, bad things began to happen. The tool could not be removed from the hatch. It just would not disengage as it was designed to do. It was stuck in place and could not be removed. This piece of hardware, probably valued at about $75, was holding up a million dollar operation! I saw the frustration on the faces of the vehicle closeout coordinator, his techs, and NASA quality inspector Johnny Corlew. They were easily tracked by color TV camera in the White Room around the crew hatch area. Had there been a reciprocal TV for them to see a view of my face that day, they would have seen a lot more frustration! What a terrible predicament! No amount of tugging, tweaking, or man-handling seemed to do a bit of good. After a long period of attempting to remove the tool, the closeout crew chief asked for permission to cut the tool handle loose and rig the hatch so we could fly. He was given the "go ahead" after a lot of discussion. Then the next gremlin sprang up. The electric saw's batteries were dead. The saw would not work. There were spare batteries, but they also would not power the saw. The firing room seemed to be in a state of helplessness. Probably thousands of shade-tree mechanics wished they could go out to the pad to help free the tool. We watched as the workers labored to remove the tool. A new threat to the launch had arisen. The winds were increasing as the morning sea breeze effect began to kick in. We were racing against degrading weather and the clock to get back on schedule.

Doug Owen was a former aircraft operations chief with a world of experience. He was Lockheed's Chief Operations boss. Lockheed was our Shuttle Processing Contractor, and we looked for them to carry the major share of the workload. Doug Owen was truly embarrassed to see his closeout crew helplessly trying to free the cheap shop tool from the expensive flight hardware hatch. Doug came to our top row location in the firing room. The normally jocular gentleman was steaming. He expressed to me and Dick Smith, our Center Director, his shame at the unfolding predicament. Lean, tall, Marv Jones, our base manager, nodded in agreement. Then Doug spoke some words I wish he would never have said.

"If I were out there, I can assure you I would get that tool out of that hatch." I'm sure, without thinking about the situation, Mr.

Smith said simply, "Well, take off. Don't stand there." Also without much thought, Marv Jones chimed in, "I'll drive you." So Marv's "blue light special" and Doug's pent-up desire to help led to a real bad day for me.

I realize now, as Shuttle Launch Director, I should have spoken out immediately and countered the direction by saying, "Mr. Smith, sir, I think going to the pad under the present conditions is not a good idea. We have man loaded the maximum number of personnel up on the Orbiter level. We should not allow additional people to go to the pad." But I didn't speak up. I don't know why. It was not my way of approaching a problem. It was not my normal disciplined approach to an emergency! I have often rethought my motive in remaining silent. The only answer I can come up with is that I was bowing to Mr. Smith's authority. This is not a good reason for such irresponsible behavior on my part! I'm sure Dick Smith, Marv Jones, and Doug Owen would agree with me completely that the four of us made a collective error in judgment. Thank God, there were no further complications of the situation. Thank God, Marv did not have a wreck during the reported ninety mile per hour race to Pad B. Thank God, we did not have to perform an emergency egress from the top of the stack, because we had too many people for the capacity of the slide wire escape baskets.

Within minutes I saw a NASA white security sedan with blue light flashing speed past one of the roadblocks to Pad B. The dynamic duo was on its way! A contractor security guard later reported that Mr. Jones' white government vehicle passed his guard post at a speed approaching ninety miles per hour! Then I saw Doug appear in the White Room, inspect the situation, and start talking to the closeout crew. By the time he arrived in the hazardous zone, the techs had secured a hand-operated hack saw and were proceeding to cut the hatch tool free. Thankfully, Marv had the presence of mind to remain at the foot of the launch structure and not further overload the design limitations of the emergency escape systems.

By the time we finally removed the hatch tool and were in a posture to launch, the winds at the Shuttle Landing Facility had exceeded the fifteen knot limits for winds blowing across the runways. Crosswinds of too high a magnitude can push the Orbiter off course and blow it sideways as it makes its final approach and rollout. They can also make nose wheel steering control difficult for the astronaut pilots. Crosswind limits must be strictly adhered to. After holding the count and waiting in hopes that the crosswinds would subside, we exceeded our launch window and were forced to scrub again. The heroics of the two speedsters, though well intended, made me feel as if I had failed to control the countdown, one of my prime responsibilities. I never received a bit of negative feedback from the dozens of test directors who reported to me. The only criticism raised was self-generated. I was not proud of that operation at all and shall always regret letting it happen. Even the media, to my knowledge, never picked up on the instance. They deserved to rake us over the coals for such undisciplined behavior. I vowed that day to never let the countdown get out of hand again. I often wanted to ask my peers and subordinates to advise me of their reaction to the added floorshow we staged that day. I like to think today that I may be oversensitive about my failure to stop their mad race to the pad. Maybe the frustration of seeing the hatch tool failure and then losing the launch to weather again caused me to be over-critical of our behavior. Nevertheless, I came to a stark realization after the Challenger tragedy on the following day! I realized that had these two fateful incidents never occurred, the entire Challenger catastrophe might also never have happened. That fateful January 28, 1986 may have been avoided had we not gotten an incorrect weather prediction on Sunday, January 26. Challenger may never have occurred had we not experienced the failure of a two-bit hatch tool on Monday, January 27. Had we been able to launch under the conditions of either of those scrubbed opportunities, Challenger's crew may have been spared. Surely the "O" rings would have sealed properly under warmer conditions and America would be relishing the lessons of a teacher in space rather than mourning seven dead heroes. What a major part every event in history seems to play. Each minute detail must be in place to lead to a significantly historic event. My soul tells me it is not happenstance. I cannot explain why things happen as they did before Challenger to thwart us on those two occasions.

I know my reaction to the day was one of disappointment. We met that afternoon to discuss weather for Tuesday, January 28. The weather folks were predicting the coldest weather for many years for the state of Florida. Also, they reported the possibility of high winds overnight. The forecast for Tuesday morning's launch time was crystal clear, visibility unlimited, chilling temperatures in the mid to high thirties and marginal winds. I really did not get alarmed at the cold forecast. We talked at length about anyone's concern for freezing temperatures over the night. I personally remembered on numerous previous occasions finding several gallons of rainwater in the External Tank intertank area. I asked the tank engineers if they had any concerns about freezing water in that area. The External Tank personnel reported that they had sealed the intertank door after getting the compartment dried and had absolutely no concerns. I thought about how we flowed cryogenic temperatures into the Shuttle elements; liquids at 300-400 degrees below zero. I knew the atmosphere outside the Orbiter in space is near -250 degrees F when not facing the direct sunlight. So my knowledge only encompassed what the space components were subjected to. I was relying on my knowledge and experience as best I knew how in examining the implications of the extreme cold temperatures predicted. I never considered the SRB segment joints or the joint O-rings to be temperature sensitive and did not ask about them at the meeting. I guess you would not call the Solid Rocket Boosters actual space hardware. To my knowledge, I had never heard the slightest reference to SRB joint O-rings not being able to seal properly at the colder temperatures. It was an accepted policy to expect those people who had a valid concern about safety or any other critical issue to report it to the highest level until they were satisfied with the outcome. This was the accepted protocol for reporting concerns in the Shuttle program. Jesse Moore, the NASA Associate Administrator for Space Flight, later discussed this reporting system before the Committee investigating the Challenger accident. "Problems and discords bubble up the line until they get resolved," said Moore. Our Flight Readiness Reviews and our management meetings just prior to launch were for the sole purpose of evaluating our status to launch. The forum for discussing concerns was there. This management meeting on the afternoon of Monday, January 27 was one such meeting. The right people were on the mission management team. All of the members of the mission management team were in attendance. No obvious constraints for a launch attempt under extremely cold weather conditions on January 28 ever surfaced. I felt good that we had addressed the hardware against the launch environment, and no one had voiced concerns. It is tragic that the SRB O-ring problem was not considered to be critical enough for "bubbling up to the top."

Of course, we did not foresee the water spillage that was to occur overnight, so we did not discuss the icicles that were to get a lot of scrutiny in the next twenty or so hours. I left for home feeling drained and disappointed, but otherwise happy to see us getting the prediction of a high front and the prospect of a clear but chilly launch morning on Tuesday. As I started to leave my office, I was handed a copy of one of the many comical writings that seem to surface at KSC in bad times as well as good times. It seems there are a lot of aspiring poets in the space business, so I only glanced at the sheet of paper and tossed it down to join the dozens of other unread documents on my desk. I'm almost reluctant to print it now, but I think that it gives some indication of how people began to feel after the long series of STS-61C launch attempts and after the frustrating day with the hatch tool failure and exceedance of crosswinds. It was intended to be sung in rapper style.

The Super Scrub Shuffle

Now we didn't come to cause no trouble
We just came to do the Super Scrub Shuffle
We missed the good weather when the hatch wouldn't close
Then we messed up when the dead drill we chose
So we sawed off the old milk stool
But tomorrow's "no go" because the weather's too cool.
We know the Launch Director could call the weather right
If we could only hold till next Friday night
First we held the Congressman, and then we held the Teacher
But holding to schedule is not our main feature.
We all hate to scrub because we know what that means
The Super Scrub Shuffle and a wave-off on beans!

...Author Unknown

I again drove home the familiar 15 miles or so and hit the bed running. The best sleep seems to come in an extended countdown when you have toiled long hours and must return again or when you have actually had a successful launch. I slept a good six hours and awoke refreshed. It was quite obvious from the chilly feeling inside the house and the struggling heat pump operation that it was extremely cold outside on this clear Florida night. As usual, I shaved, showered, and dressed in my warmest suit to attack the freezing night air. It seemed that my old second car, reliable as she was, would never get warm enough to keep my teeth from chattering as I drove north to KSC. It was a night filled with a million bright stars, an astronomer's dream. As I passed the first guard shack that was not manned, I mused over how the blades of an idle electric fan were turning around at a fairly fast speed propelled by the strong night winds. These were the same wind components that were playing havoc with our makeshift freeze protection at the pad at the very time I was driving to the Launch Control Center. My thoughts were on 8:38 a.m., Tuesday, January 28, 1986, our projected T-0 for Challenger's next launch attempt.

CHAPTER 19

The Evening before Launch

For by wise counsel you will wage your own war,
And in a multitude of counselors there is safety.

Proverbs 24:6

A lot of folks who have visited the Launch Director's Office in the Launch Control Center are impressed with its location and unobstructed view of the two Shuttle launch pads. Very small by executive office standards, it is a popular spot known for its splendid view rather than its ambiance. Located on the fourth floor adjacent to the ceiling of Firing Room 3, the room is always a beehive of activity. The manager of Launch and Landing Operations was a key government position at the Kennedy Space Center. KSC's processing, scheduling, manifest support, and operations management were focused under the leader of this important office. In 1986 the position also included the responsibility of Launch Director. In launch countdown, from about Launch minus 3 hours, the very close members of the astronaut families gather in this office to wait excitedly for a chance to see their loved ones soar into space. Escorted by a KSC Public Affairs representative and a NASA astronaut, the families are treated royally. We provided treats for the astronaut children and refreshments of every kind for the astronaut spouses. My secretary, Barbara Rabren, was especially good with the smaller children of the astronauts, and they all seemed to attach themselves to her immediately. We provided coffee, soft drinks, donuts, cookies, and even toys for the smaller children. A launch tradition had also evolved with the school-age youngsters and teenagers in the families. We asked these older kids to color a commemorative "picture" for their dad or mom's launch using Magic Markers and a three foot by five foot piece of white paper which we provided. These sharp innovative youngsters have provided some excellent space art which may one day be displayed on a space museum wall somewhere in America. Every painting portrays the applicable crew's mission patch and a lot of "hip, hip, hoorays" for their favorite astronauts. We framed these one-of-a-kind creations and hung them on display on the walls of the LCC's fourth floor. As we fly more and more Shuttle missions, we will have to find more wall space or build an exhibition wing onto the launch center. This artwork was a fascinating activity that we certainly planned to continue as it leaves a lasting memory of the time families spent near the launch facility.

When I arrived at the LCC on the evening of January 27, I parked in the very nondescript Launch Director's parking place provided on the eastern side of the building. This spot provided easy access to the south elevator which opened in front of the Launch Director's office on the fourth floor. When I arrived at my office about 11 p.m., I found a late evening teleconference in progress involving Stan Reinartz, the Marshall Shuttle Projects Manager, Arnie Aldrich, the Shuttle Level II Manager, and my representative, Gene Sestile. I assumed Stan and Arnie were talking from their motel rooms. I had decided with so much launch activity in the last year to assign a senior manager to cover the night shift during countdowns. I wanted a senior NASA person who would be trusted to keep on top of all major issues. Gene Sestile was a tried and true veteran who knew his way around. He had a lot of experience in hardware, testing, and manifesting. I felt comfortable with Gene representing Launch and Landing Operations in any forum at any level. The teleconference was very important. The winds over the Atlantic near the SRB recovery area were tremendously high. The two SRB recovery ships were having trouble in the high seas off Cape Canaveral with waves approaching 25 feet. The captain had radioed to shore asking for permission to come closer to land as he feared for the safety of the ships' crews. He had already returned 25 miles closer to shore from the normal station-keeping distance for recovery operations. Reinartz was advising us and Arnie Aldrich that if we abandoned the recovery area much farther, we would risk losing the spent solid rocket boosters after they splashed into the ocean the next morning. If we held the ships in the high seas, we risked the lives of the seamen and loss of one or both vessels. It sounded like a pretty easy decision; actually it was not! We were certainly not going to jeopardize the recovery ship crews. We also wanted to recover the boosters. There were over 20 million dollars in flight equipment that should be recovered and refurbished. We did not want to lose our launch opportunity on Tuesday. We considered that the seas were predicted to be twelve feet tomorrow. We knew we had fairly reliable recovery beacons on the boosters so we could probably locate them easily. Gene and I contributed very little to the rationale discussion except to assure Arnie that the safety of our seamen was a top priority. We emphasized that we wanted the ships to get into a safe posture even if it meant returning all the way back to port. So Arnie decided, I think quite correctly, to give the ships the leverage to come back westward to where they felt safe and no longer in jeopardy. This allowed them to return to port if they saw fit. We also decided to continue the countdown toward launching the next day. I think all three parties involved felt that our recovery chances were good, and there was a low probability of losing the rocket boosters. During the teleconference, no reference was made to any concerns about SRB "O" rings or super cold temperatures. During the investigations after Challenger, I was to learn that Arnie, like my organization, was not made aware of Marshall and Thiokol's discussions and lengthy teleconferences regarding SRB joint O-ring seals. I'm also certain that Mr. Reinartz did not see the need to further elevate the O-ring concerns. He surely felt that it had been argued and rationalized at the highest necessary level of management and "put to bed." We ended the teleconference with full agreement to direct our ships to seek safe waters and our launch team to continue to prepare to launch!

As I grabbed my operational communications headset, my logbook, and my countdown procedures to go to the Firing Room, I glanced at the beautiful scene out my office window. My office was prime viewing space for the launch pads on a clear night, second only to the two main Firing Rooms where only launch team members are allowed. The pad is spectacular at night during countdown with the golden external tank and the pure white Orbiter illuminated in splendor by dozens of super-bright Xenon lamps. The brilliant lamps, powered by motor generator sets, give clear visibility for all the crooks and crannies of the pad structure. The operational TV camera depends on these lights to allow minute area viewing. The bright aura around the pad caused by a halo of light added a Buck Rogers air of magic. As the external tank is filled with thousands of gallons of liquid oxygen and liquid hydrogen, one begins to see the bright red glow from the hydrogen burn stack and to notice the clean white plume of the LOX venting off through the LOX beanie cap at a periodic rate. The mere fact that the pad is deserted by humans for the hazardous tanking phase seems to add to the serenity and technological splendor. Of course, the night critters are out there; bugs, birds, mosquitoes, small animals, who knows what else? Often a glance out the big windows can be mesmerizing, and you have to force your mind back on track again.

I hurriedly walked down the two flights to my seat in the top row of the freezing confines of Firing Room 3. I carefully and comfortably donned the Launch Director's lightweight headset which would then become a part of my body for the next twelve or more hours. I adjusted the channel selections and volume controls on my Launch Director's Operational Intercommunications Unit. It was good to hear familiar male and female voices conducting the Shuttle countdown with sharp precision and genuine professionalism. There is never any clowning around on the voice nets. Space launch is serious business, and the launch team always performs in a skilled, professional, and disciplined manner.

Immediately upon reporting that the Launch Director had checked in on the net, I was briefed by the Test Directors, the Project Engineers, and the Flow Director on the status of the countdown. The report was not encouraging. The weather outside was exceptionally cold, the temperature in the low thirties and continuing to fall. The winds were extremely high, but were expected to decrease before dawn and be within our Launch Commit Criteria (LCC) limits by launch time. But our biggest problems were more apparent, something to be reckoned with now. There were four significant problems on the pad, and I immediately set the planning into place to get them corrected before we would attempt to launch Challenger. My crackerjack launch support team members were way ahead of me, as expected, and my only decision was to assure them that we would bring each of these safety critical problems to closure. None of the problems merited a lot of consideration for waiver; they were all mandatory for launch in our opinion. I made it quite clear over the voice loops that the Launch Director considered the problems as being constraints to launch. We didn't spend a lot of time discussing them; we went to work. As I settled into the normal work of Launch Director in the Shuttle Firing Room, I was almost stunned by how unmercifully cold it was at my seat. Not only was the cool air blowing through the consoles to keep the electronic hardware cool, but the outside temperature plummeting near freezing added to the misery. The thick plate glass windows seemed to transport the outside cold directly to the back of my neck and head. The cold console air took care of freezing my hands and feet. There's no such request as "Hey, turn the heat up in here!" Despite the cold, I decided that the gravity of the four pad problems was important enough that I stay at my position throughout the night. I did not leave the firing room except for the two or three times for necessary breaks. I left the coffee breaks off although coffee may have helped break the body chill. The four problems demanded my total attention. I decided that the off-line meetings conducted by the program managers were for their decision-making. I felt my best service could be done monitoring the progress of the actions we were taking to correct our problems. The problems were:

1. When we cleared the pad for tanking, we left several water hoses dripping in our pad water system to prevent freezing. It is a commonly used method of freeze protection by some who live in cold regions of the country. We had decided earlier to leave the pad with water slowly running. But during the pad clear period that evening, the high winds had dislodged one or more of the water hoses. Ice had formed over a lot of the service structure. Icicles as

long as two feet could be seen hanging from the structures on the operational television. We sent a red crew into the pad to evaluate the ice situation and take whatever corrective action deemed necessary.

2. Underneath the SRB flame trenches, we had installed a series of water bags intended to diminish the shock effect caused by the SRB flames at T-0. This eliminated a shock wave of sound rebounding from the flame trench and damaging the Orbiter aft structure and the SRB skirt area. The water bags actually contained a small amount of glycol, but it was not enough to keep the water in these bags from freezing. A slush had formed and was beginning to freeze more solidly. The glycol mixture in the bags of this sound suppression system was needed for launch, and we could not launch with it as a "slush" rather than a liquid. We sent special experts into the pad area to determine a solution to this dilemma.

3. One of the Hardware Interface Modules, appropriately called a HIM, had failed. These electronic units, located inside the mobile launch platform, were used to collect a multitude of ground processing data. But more importantly, these units were the interface that allowed the launch team console engineers to command and control the thousands of functions that must be operated to facilitate a Shuttle launch. This particular failed HIM unit was one of the ones that collected fire detection data and without it being operational, thirty of our fire detectors were reading erroneously.

4. Several of the fire detectors on the crew's emergency egress route were not functioning properly. I determined without a lot of thought that we must have these detectors operational prior to crew ingress.

I distinctly remember discussing each of these problems over the communications nets. I declared that each of these problems was considered a constraint to launch by the Launch Director, and we were to proceed to "fix" them prior to launch. We actually held up tanking of the External Tank super-cooled cryogenic liquids for two hours as the emergency red crews worked to evaluate and fix the four pad problems. We had as many people popping off icicles and probing on water bags with broomsticks as we could allow on the pad. We added more glycol to the water solution in the bags. We discussed for hours the possible impingement of icicles on the

Orbiter tiles at lift-off. The Rockwell Downey support team ran estimates and simulations to try to formulate the impact of any stray ice on the pad. There were a lot of meetings held off-net by Arnie Aldrich and his element project managers. There were telephone conversations and on-net discussions. I still elected to stay where my assigned responsibility was most needed. I can only describe the work of the red crews, the ICE inspection team, and the engineering analysis support teams as totally professional, responsible in the most dedicated way.

On one of my short restroom breaks, I met Horace Lamberth, our KSC Engineering Chief, and an assistant, Jose Garcia, in the hallway. We discussed the rationale of whether we should proceed to load the External Tank. After our short discussion, I saw no reason not to proceed to load the LOX and the LH2 into the External Tank. It was my responsibility to make the call and rightly so. My rationale for proceeding to tank was built around there being no risk with flowing the super-cool cryogenic liquids in an extremely cold outside environment. I felt that we could go ahead with tanking and discuss our launch options when the remaining vehicle systems experts and the rest of our mission management team reported on station. Looking back on my decision to tank that night, I can find no reason that this was a bad decision. One would have to come up with some very illogical reasoning to convince himself that we should not have tanked. All this reasoning would have been subjective and not supported by the technical information available to us at the time. Although they were easier to address, the fire detector problem and the HIM problem were also handled as constraints to launch and both required sending red crews onto the pad to correct.

Much time and effort were focused on correcting these problems all through the night. I opted not to attend the numerous meetings regarding ice, temperatures, Launch Commit Criteria, redlines, waivers, icicle trajectories, water bag requirements, etc. I knew that our best managers like Bob Sieck were on top of these issues. I relied heavily on program management, primarily Arnie Aldrich and Jesse Moore, to advise me of any technical constraints to launch. This is in no way an attempt to pass the buck. It was our established way of doing business, and I think it was effective. The very best space launch practitioners in the world were practicing their trade. The Rogers Commission was later to label it as being "flawed." I do not believe a fool-proof system of decision making and reporting can exist without some margin for error, especially for a huge, multi-faceted complex operation like the Space Shuttle Program! I again feel that a judgment call was made that affected history and resulted in the death of seven of our greatest. It is sad that a system can allow for internal destruction due to a breakdown in discipline and an erroneous assumption that the O-ring problem had been sufficiently debated at the proper levels. To my knowledge, after the final Marshall-Thiokol O-ring teleconference ended around 11 p.m. on the eve of the Challenger accident, it was not discussed again in any of the subsequent constraints meetings held during the night.

I recall one disheartening reference after another by the press who were quick to voice opinions regarding why we had the audacity to launch on such a cold day. "Why did they launch when it was so cold?" "Didn't the icicles indicate it was too cold?" "What were those people thinking about?"

I can assure all doubters that we were long deliberating on the cold. We spent hours studying and destroying the threatening icicles. As Launch Director you are never in a position to ask: "If it's cold, how cold must it get before you scrub?" It is not a simple "too cold to launch" answer. The launch team must have a launch temperature limit. Our Launch Commit Criteria set that lower limit redline at 31 degrees Fahrenheit.

Some critics have even compared the Challenger "cold launch" decision to the sinking of the Titanic. Reportedly, the crew of the Titanic was warned of the ice conditions on that maiden voyage and yet ignored it. We did not ignore the ice formations! I am certain should Challenger have flown its STS-51L mission successfully and returned to Kennedy safely, we would have found no tile or Orbiter structural damage caused by the ice that formed on the pad before launch! We would probably have received plaudits from the media for the splendid way we worked all our pre-launch constraints and managed to launch in spite of the low temperature!

As the clear dawn began to break over the Atlantic, the bitter cold of the outside air and the equipment cooling inside the Firing Room combined to make my seat feel as if I was in a giant frozen food locker. The outside temperature had finally dropped to a low of 28 degrees F, but the wind chill factor would have placed it in the low teens. Still the red crews and the ICE Inspection Team worked tirelessly to break icicles, add glycol to water troughs, evaluate the icing conditions, and eliminate the slush in the sound suppression water bags. The glowing eye of the bright golden sun slowly peeping over the ocean waves far, far away gave me hope that with the sun's rays the ice would begin to melt. Our new predicted launch time was now closer to eleven o'clock. I was visited at the Launch Director's chair several times during the night by Arnie Aldrich. He advised me of where their analysis was regarding the icicle concerns and other problems he was tracking. It was quite obvious Arnie was totally immersed in the launch decision activities. There have been no truly professional managers in NASA better than Arnie Aldrich. He not only possesses an intelligent mind, but wisdom to sort out the intangibles. He is one who can separate the chaff from the wheat. I felt extremely confident in Arnie's reports of what was going on outside the Firing Room as the launch team proceeded. Arnie never mentioned any problems with booster O-rings and I'm convinced that he, like I, never was advised of the existence or the severity of the situation. I also remember distinctly a visit from Jesse Moore after he had arrived in the Firing Room. His normal gentle smile expressed no strong launch concerns. Jesse had made his informal off-net poll of his Shuttle managers across NASA, and he voiced his support to proceed to launch!

The executive summary of the Rogers Commission's investigation of the Challenger accident was succinct, but I think it well described the events of the evening and night activities: "The weather was forecast to be clear and very cold, with temperatures dropping into the low twenties overnight. The management team directed engineers to assess the possible effects of temperature on the launch. No critical issues were identified to management officials, and while evaluation continued, it was decided to proceed with the countdown and fueling of the External Tank."

One of the first things I decided to do the morning of January 29, 1986, after the terrible disaster of Challenger the day before, was to document as accurately as I could recall the events of the countdown. I knew I must record as best I could remember the salient happenings of that night. I did not recall exactly what I had written in the Launch Director's log, and it was impounded as part of the investigation. I never saw my log again after January 28 and assume it was stored with the other official Challenger records. The only inconsistency I have found in my written post-incident report was the fact that the voice on the teleconference regarding SRB-recovery ships was actually that of Stan Reinartz of Marshall rather than Larry Mulloy. I also only listed three of the four problems we were faced with on the night before launch. My purpose for documenting my abbreviated account of the events was merely to record my thoughts prior to the passing of too much time. I also wanted to document what I thought was the process we used to address our problems and reach closure on them. My written words, with no input or review by another individual, were exactly as follows:

"On reporting for launch late 27 January 86 (approximately 11 p.m.), I found a teleconference concluding in the Launch Director's office. Included on the teleconference were Gene Sestile of my staff (2nd shift management rep), Larry Mulloy, MSFC SRB Project Manager, and Arnie Aldrich, JSC STS Level II Program Manager. The purpose of the teleconference was to assess the fact that the SRB recovery vessels were experiencing 25-30 foot seas in the recovery area approximately 100 miles east of KSC. The commander had determined the condition an emergency, and the safety of the crews was uncertain. He ordered the ships to proceed toward shore about 25 miles, and they were at that time station-keeping in 12 knot winds. The feeling from the teleconference was that we would lose the frustrums and the parachutes (approximately 1 million dollars) but not the SRBs. With the concurrence of Mulloy and Aldrich, we accepted this as program direction to proceed with launch.

I reported on station in Firing Room 3 around 11:30 p.m., 27 January 1986 and at approximately 1:15 a.m., 28 January three problems surfaced at Pad B affecting the start of main cryo tanking:

1. HIM 5985 was inoperative - this disabled Fire Detectors 1-30 - these fire detectors are mandatory for tanking.

2. Several fire detectors on the Orbiter Access Arm (OAA) were not operating - needed for crew emergency egress warning.

3. A water leak had developed at the 220' level of the Fixed Service Structure (FSS) - a hose used for diverting water for freeze proofing had been dislodged - located approximately 65' from the Orbiter but producing considerable ice on the FSS.

All three problems were considered constraints to ET tanking and a red crew was sent in for each problem with a hold placed on start of tanking. We basically lost 1-1.5 hours because of those 3 problems. The HIM repair was the time driver. We changed several cards before finally changing the faulty one. The water leak caused icicles mainly on the northwest side of the Fixed Service Structure (FSS). Some of the icicles were as long as 18 inches. The red crew redirected the H20 hose leak and did little else except assess the ice and its locations.

When the ICE Inspection Crew returned to the LCC, we (Thomas, Lamberth, Stevenson) held a short off-net meeting. Lamberth and Stevenson felt the FSS ice was manageable, but they expressed concern for the icing in the SRB pressure suppression water bags. We held the start of loading for 10-15 minutes to discuss these conditions and I (as Launch Director) made the decision to start tanking with the concurrence of Engineering.

We felt that we would work the ice problems when more of our experts and management reported on station. I made this decision based upon no LCC for icing but remembered the 31 degrees F lower limit for launch. THIS WAS A DECISION TO TANK, NOT TO LAUNCH.

While main cryo tanking continued with few additional problems, several meetings were held off-net to decide on the approach to the icing situation. I did not attend these meetings, relying on engineering judgment on these calls. Engineering kept me advised periodically of their progress.

Based on a recommendation from engineering, we sent 2-3 extra facilities personnel to the Pad with the ICE team at the T-3 hour normal ICE Inspection time. The team reported approximately ½ inches of ice on most of the ice bags in the SRB troughs. The ICE team spent a good bit of time breaking up this ice. They were able to dip out approximately 95% of the ice from the bags.

The ICE team leader reported on the OIS that the tank ice was nominal (that is, no abnormal tank icing), no leaks, nominal ice acreage, and that the external tank condition was as good in general as it ever had been. He also reported the status of the icing on the FSS as basically the same conditions. He reported that there was only "slush" remaining in the water bags.

The ICE team and several of our managers continued to meet to discuss further actions required while the countdown continued through crew ingress. I was asked again late in the count (L-1 hr. time frame) to allow a small ICE crew back on the Pad to remove ice from the "0" level of the MLP. The crew went to the Pad surface and swept all free ice from the surface into the flame trenches. The management advised me that they no longer felt that the FSS ice was a threat since their calculations did not show the ice would travel across to the Orbiter. They felt the ice bags were no longer a hazard, and the free ice had been removed that would have been a debris source. The concern was then one of waiting too late until the ice began to "thaw loose." We then set 11:38 EST as desired T-0 and began to press toward that time.

At no time did I have any idea of cold temperature affecting SRB joint (O-ring) performance. My guidelines were strictly from the LCC of 31 degrees F to 99 degrees F as launch temperature redlines. I pulled on several ideas by asking about certain heaters, about ice in the ET intertank, etc., and all my expertise failed to report any critical temperature concerns. I felt perfectly at ease about launching with the temperature in the high 30's, and once we cleared the "0" level ice, I never again thought about the cold, concentrating on the technical aspects of the launch countdown itself."

There have only been one or two Shuttle launches in which the Launch Team and management have not worked late countdown problems. Some of the late decisions have backfired. A Shuttle Spacelab mission in April 1997 on Columbia was shortened to four days. The crew performed an early return from orbit after an onboard fuel cell had to be shut down. The launch team was confronted with symptoms of a fuel cell failure late in the countdown, but decided Columbia was ready for launch.

Some authors who have written about the Challenger launch decision have implied impropriety, negligence, and poor judgment by the Shuttle launch team. As its leader that morning, I can relate the facts - I know the truth!

1. There was absolutely no negligence by any member of the dedicated Shuttle launch team.
2. There were no blind drives or ambitions; no hidden agendas.
3. There was no pressure from anywhere to launch.
4. There were no madmen at large, no evil scientist, and no Frankenstein monster.
5. There were no sinister plots unfolding.
6. There was no disregard for the value of life.

Assuming there had been no O-ring concerns, given the same set of launch conditions (icicles, low temperatures, sensor failures, etc.), I am certain my decision to launch would have been the same. Were I asked to make the launch decision 1000 additional times, I would reach the same decision 1000 times.

What is most perplexing and mind-boggling is the question raised from the testimony given by Larry Mulloy before the Rogers Commission regarding passing up the management chain the concerns about the effects of low temperature on the SRB joint O-rings:

> "In my opinion, had the same information...been passed along by Arnie Aldrich and Jesse Moore, the judgment to proceed with the launch would have prevailed."

Would Arnie Aldrich and Jesse Moore have given me as the Challenger Launch Director a recommendation to launch had they heard the full O-ring story? I certainly want to trust in the integrity of the communications and control management system we had in place at that time. I want to believe that a lot of higher level decision makers would have at least scrubbed the launch for that day and taken time to listen to the objections presented to Marshall Management by Thiokol. We will never know the answer to this intriguing "What if?" question!

CHAPTER 20

An Ever Present Danger

My job as Launch Director for the Shuttle was simply a progression of career assignments from one job of increasing importance to another. It felt quite natural, after serving in two other highly responsible firing room positions, that I assume the most responsible position of Launch Director. I had totally enjoyed my time spent as an Orbiter Project Engineer and as the Shuttle Project Engineer. As they might say in a corporate boardroom, I had "served my time."

The launch of Challenger STS-51L was to be my fifth as Launch Director. I had served as Assistant Launch Director for two other launches. I had reached the point of comfort, occupying the critical Launch Director's position. After my first three launches were picture-perfect, Columbia's January 12 launch had followed a long struggle which involved six scrubs. At the post-launch press conference, Bob Sieck and I looked tired and drained. Despite the dark circles under my eyes, I felt relaxed and confident in responding to the questions from the media. These were hard times, but these were good times!

What does the Shuttle Launch Director do? He sits on the top row of several rows of launch management consoles in the Launch Control Center, the LCC. The LCC is five or six miles west of the Shuttle Pads A and B. These two giant pads are located only 200-300 yards from the roaring surf of the Atlantic Ocean. The pads rest amid scrub palmetto brush and sparse ground cover, surrounded on several sides by marsh and lagoons. This beachside setting would be ideal for the location of plush ocean resorts that are so familiar along both Florida coastlines. From the Launch Director's seat, one can clearly see the two pads through the gigantic plate glass windows three feet behind the console. Below on the firing room's main floor are located twenty or more consoles each manned by ten to fifteen engineers. These people are the experts for all of the Shuttle systems and the associated ground support equipment required for launch. As expected on launch day, our launch team looks a great deal more formal. Ties are in order for most of the men and chic outfits for the women. The Launch Director console is well-equipped with operational voice communications, color television monitors, telemetry screen, weather displays, binoculars, and peppermints for stress relief! By pushing buttons, the Launch Director can select to talk or listen to hundreds of engineers and advisors around the world. This expertise includes safety specialists, test directors, managers, project engineers, systems specialists, test support managers, weather experts, flight directors, range safety officers, computer operators; a host of experts from every imaginable field of endeavor. This outstanding group of men and women constitute the Shuttle launch team whose motto is "Doing what Others Dream."

The Launch Director is the leader on launch day. Like most critical job positions, this person is not the initiator of a lot of decision options, only the deciding vote maker. On launch day, the buck stops at the Launch Director's post. He makes a lot of crucial decisions; he controls the clock; he judges weather data as to whether to proceed to fill the giant External Tank with liquid oxygen and liquid hydrogen. He concurs with waking the flight crew on the morning of launch. He determines if a "red crew" should be allowed to enter the hazardous pad area to carry out a task that is off nominal. He accumulates the expert advice from his subordinates, peers, and those who manage the Shuttle program. He is in no way a unilateral decision maker. He must weigh the advice of many, often choosing from opposing opinions. He is surrounded by the cream-of-the-crop, tried and true veterans of space launches over three decades. He listens to six or eight voices in his operational intercommunications headset and learns to differentiate between important reports and routine ones. He is supported by people at Transatlantic Abort sites in Spain and Africa, at emergency landing sites all over the globe, by the back-up firing room's cadre of experts, at vendor plants, by ships at sea and planes in the skies, by other NASA centers such as the Johnson and Marshall Space Flight Centers, at range-tracking stations, and at weather stations. There are helicopters deployed for emergencies, even army surplus armored vehicles to rescue the astronauts in a ground emergency. In concert with the Flight Director at the Johnson Space Center, the two team leaders have the final responsibility to commit the space vehicle to launch.

On launch day, the Launch Director is the quarterback. People look to him for leadership, stability, quick-thinking and a cool demeanor. I think that the most important characteristic a man or woman can bring to the Launch Director's job is respect. The Launch Director must have the respect of his launch team, and he must respect each and every member of the team. I personally feel that the stress and complexity of the Launch Director position is exaggerated. I found the stress to be far less than what I had expected. However, the job can be physically and mentally demanding. Teamwork is paramount! The job can and should be enjoyable. Each launch is like a Super Bowl. One has sacrificed to get there, and he wants to wear that big ring. There is a great team, and they love to play ball. The team supports you and each other. There is always a euphoric feeling of the total success when you see the huge powerful Shuttle pounding the atmosphere as it rises majestically into the skies. The Launch Director's job is rewarding. It must not be underestimated in its importance, however. One mistake can cause a catastrophe! One short cut can lead to disaster!

The launch team at the Kennedy Space Center respects and works hard for the astronaut flight crew and the Launch Director. Everything is a team effort with no prima donnas. I have reported to a firing room on many occasions to be met by a smile, a wink, a nod or some similar body gesture and immediately felt assured that all was well! These men and women are dedicated guardians of one of America's premier national assets. They do not take their responsibilities lightly.

The Firing Room atmosphere is not Hollywood. There is little, if any, glamour in these acting roles. It is pure professionalism. It is not high drama, but still it is the best of what Americans can be proud of and brag about. It is not the made-for-the-screen theatrics of *E. T.* or *Alien.* And it is centuries from being *Star Wars.* It is real people in real situations. It is a real triumph of man and machine. Launching Shuttle crews on the best journey in the universe has been called an E-ticket ride. Not everyone remembers the early days of Disney World when one purchased a ticket book for a day's visit. The ticket book usually included two or three E-tickets, the ones that got you the most for your money, the best rides in the park. The Shuttle launch team still gives E-tickets.

Sitting in the Launch Director's chair is exhilarating! It is exciting! It is gratifying! To be or to have been a part of such a masterful operation of such a high technological nature is something that only a few have had the privilege to experience. I am eternally grateful to have been the Shuttle Launch Director for slightly more than a year. My last time to serve as Launch Director was the fateful Challenger mission. I do not understand the master plan of the Creator who I am certain predestined that I would be in that position at that time in my journey through life. That is how I feel and what I believe. I expect to find an eternal answer to the question that man has asked through the ages, "Why me, Lord?"

What a spectacular sight the Launch Director has of the great white structure poised on its launch pad, waiting to carry its riders beyond the grasp of earth's gravitational bonds. From atop the Shuttle's huge External Tank, a pure white vapor of gaseous oxygen is vented periodically to relieve pressure in the tank. Its powerful twin in the combustion process, gaseous hydrogen, presents an eerie torch to the clean sky as a bright orange flame is produced at the tip of the hydrogen flare stack. If one has ever driven past a huge oil refinery at night and sensed the mystery of the flares and vapors, it is easier to sense the appearance of this pre-dawn Shuttle sitting on its launch pad. What a fantastic combination of gas, flame, lights, reflections, hisses, creaks, and similar sounds that create a Buck Rogers atmosphere around the great mechanical structure of the Shuttle. Man will mount this enormous chariot of fire and ride it into space!

Once the giant External Tank has been loaded to the maximum with liquid oxygen and liquid hydrogen, the Test Directors open up the pad for limited access by special work crews. On a nominal countdown scenario, red crew emergency entry to the pad is not required. The two non-flight teams who routinely go to the launch pad are there to perform essential pre-launch work. We send a closeout crew comprised of a chief, a NASA quality specialist, two or three Orbiter closeout technicians, and an ASP (Astronaut Support Pilot). This acronym was not chosen because it describes the personality of some of those who have held the job. The ASP is the key to getting a crew successfully on board, seated, and secured. Congenial Loren Shriver served on STS-1 as our first ASP. I recall how we had a breathing air hose disconnect from Commander Young's seat after he was ingressed. Thanks to Loren's magical manipulations, the hose was secured in a blind spot, connected only by touch and feel. Working as a contortionist would, Loren managed to reconnect the life critical air supply. This last minute work resulted in a successful launch of the STS-1 mission. The only time we disappoint an ASP is when we refuse to allow one to get a Shuttle flight as a stowaway, something NASA has been successful in avoiding so far.

The other ground team going to the pad after tanking is the ICE team. I hesitate to even try to determine the origin of the acronym ICE. My memory says it means Icing Condition Evaluation (ICE). The team's function is to scan the flight elements of the Shuttle stack for ice buildup caused by the super cool fluids we load on the Shuttle and affected by the atmospheric conditions of the day. At approximately T-3 hours, the ICE team, led early in the program by Charlie Stevens and Greg Katnik of NASA, boards a government-owned van and departs for the Shuttle pad. At the same time, the Orbiter closeout crew is cleared to enter the hazardous pad area. When the External Tank has been filled with super-cool hydrogen and oxygen, ice tends to form in certain areas. If ice forms in the wrong areas, the tremendous energy of a Shuttle launch could dislodge the ice and cause it to impact the Orbiter's fragile tiles. Clad in bright orange fire protection coveralls, the ICE team is quite colorful and easily recognized by ground controllers watching on color television monitors. Each crew member has an identification number for tracking purposes. The numbers are prominent for obvious reasons; over the heart, both arms, and on the back. The team describes the Shuttle stack as being "alive." Katnik compares the purging and venting of gases like the sound of a runner hyper-ventilating. Working in two-man teams, the ICE team scans all the critical areas of the Orbiter, the main engines, the solid rocket boosters and the external tank, for build-up of ice that could constitute a potential cause of debris damage. They have two hours to complete their review and gather thermal data. They use infrared scanners and thermal imagers to determine the temperatures and cold spots anywhere on the flight element hardware. Ice formation is their main concern. The Shuttle Launch Commit Criteria rules do not permit launch if ice forms on the tank that is thicker than 1/16th inches. In the history of the Shuttle launches, not a single launch has been scrubbed due to ice formation being too excessive. The team's judgment has always been recognized as credible by the launch management. Should the ICE team ever recommend a hold or scrub because of ice build-up, their knowledge and experience would not be questioned. This small professional team is a vital contributor to every successful Shuttle mission. The ICE team played a significant role in determining that the icicles which formed on the pad before the Challenger launch had been cleared sufficiently to launch. They served as the eyes and ears for a lot of us who relied on their expertise to make the correct judgment call on the impact of the ice accumulation.

At about T-3 hours and 30 minutes, the Shuttle flight crew was awakened, had showered and eaten breakfast. They had ceremoniously acknowledged the traditional cake bearing a reproduction of their mission patch in sugary colors of icings. Tradition among astronaut crews is never to cut and eat even a bite of the cake! This practice has passed along to all crews; most members never even ask what became of the celebratory symbol. Breakfast is never a priority for most Shuttle astronauts. Some say that the best posture for launch is on an empty stomach.

A far cry from the bulky self-contained, sealed pressure suits of earlier manned programs, the orange launch and entry flight suits are very similar to the flight suits worn by so many military aviators today who fly high performance jet aircraft. No longer do our crews have to carry the burdensome air packs to maintain suit integrity and to keep them cool enroute to the pad. After a suit integrity check and test of the crew's helmet and communications in the suit-up lab at the Operations and Checkout Building, they ride the elevator three floors down to board the converted Airstream van. This big silver aluminum van transports the crew from their temporary quarters to the Shuttle pad in comfort. Even after scores of launches, an enthusiastic band of media, friends, and KSC employees bid farewell to the crews as they board the Astrovan for the final few land miles of their fantastic journey.

This scenario is very similar to the series of events we followed on the day of the Challenger launch.

Back in the firing room, systems have been activated in a disciplined order. Fuel cells are being purged periodically to prevent the buildup of contaminants on the electricity producing components. The hazardous gas sensors sample the Shuttle closed compartments for any leaks. The ascent flight team under the leadership of Ascent Flight Director Jay Greene has reported on station at the Johnson Space Center's control room. Through our pre-launch association prior to Challenger, Jay and I became real close friends.

We once laughed at how strange it was that the Flight Director Jay Greene, of Jewish descent, was married to a beautiful Hispanic lady named Mary. And the Launch Director, Gene Thomas, a Christian believer, was married to a beautiful southern belle named Juanita. Jay and I, both being avid readers, often recommended good books to each another. It was not at all surprising to get a short phone call from Jay saying, "This is Greene; read so and so." I often placed calls with similar information to Jay. The dedicated people of Shuttle are exemplified by so many outstanding ones as Jay Greene.

Capsule Communicator Dick Covey was anxiously waiting to communicate with the Challenger crew. The Orbiter's Inertial Measurement Units were receiving their final alignment checks. The Ground Launch Sequencer was sitting on ready to perform its critical launch functions. On the night before the Challenger accident, a lot of extra activity was being managed. We were doing yeoman duty in correcting the four major pad problems that we were facing. As the night ended and the cold bright day approached, we had held our countdown two additional hours to fix these problems. We had been projecting a launch time of 9:38 a.m., but the holds for two hours pushed us out to 11:38 a.m. I was in no way disappointed in delays when we were in the process of fixing our problems. I think our determination to correct the pad problems is a testimony to our emphasis on safety. We sent an ICE evaluation team to the pad during the hold at T-20 minutes to get a final look at the icing conditions. We were confident we were ready to fly when we committed to do so.

What an engineering marvel is this mixture of Herculean space machine, complex technical ground launch equipment, and well-trained, disciplined team of launch experts. The Space Shuttle Team: Doing What Others Dream! Our ground computers, either Hezekiah in Firing Room 1 or Methuselah in Firing Room 3, communicate from the Firing Room to the Orbiter's four on-board General Purpose Computers at a rate of one million bits per second over five miles of cable or fiber optics. In the early 70's, a one megabit data bus was fast. In today's world of high rate information technology, it is a Model T Ford.

In one of my speeches, I used this information about the speed at which the ground and flight computers communicate to emphasize a point. After the speech, a small spectacled man came up to say hello. He commented, "If you think that data bus on the Shuttle communicates fast, you should hear my wife and her girlfriend talking on the phone."

A major key to the success of the launch countdown rests with the software packages loaded in the Ground Launch Sequencer (GLS). The GLS controls critical commands to the Shuttle vehicle and the ground support equipment. It conducts all critical data checks, countdown "hold" management, and the important function of the countdown clock. The Launch Director, through the Test Directors, tells the GLS operator what the desired launch time is to be based upon mission data and parameters provided by mission planners. Most important of this data is the predetermined launch window, usually dictated by the nature of the mission and the desired orbit of the payloads. A rendezvous and docking mission normally has a very short window because the Orbiter wishes to launch such that its orbit does not differ significantly from the orbiting object to which it is to rendezvous. Happiness for a Launch Director is having a long launch window with which to work. Often the only launch window restrictions are medical limits on allowing the astronaut crew to lie on their backs for an excessive length of time. NASA has a Launch Commit guideline of allowing the crew no more than two and one-half hours in the launch seated position once they ingress their seats. Most of my astronaut friends have told me that this time is plenty enough to lie on your back, feet and knees above the line of your spine, strapped in a metal chair. Sounds almost like some medieval torture chamber!

Once the GLS operator sets the prescribed launch time as instructed by the Launch Director, if no one calls a hold or stops the clock, the Launch T-0 will occur within 10-30 milliseconds of the desired time. A millisecond is, to the man on the street, one thousandth of a second. The GLS controls the critical functions of the countdown from T-9 minutes until T-0. At T-20 minutes, the onboard computers are transitioned from a pre-launch software configuration to an ascent software configuration. After the clock resumes counting at T-20 minutes, you begin to get a confident feeling that launch is imminent. The crew is secured inside the Orbiter cabin, the cabin and hatch pressure integrity are verified, and the communications links are established. The closeout crew has cleared the pad area. The range has verified its command link that can destruct the Shuttle should it veer drastically off a nominal ascent trajectory and endanger inhabitants and cities on the ground. The weather reconnaissance aircraft, usually one of the Shuttle trainers, a modified Grumman jet, has given a prediction of weather to be expected over the launch site at T-0. The weather plane flies an area 50-60 miles around the KSC to survey pending weather, either that coming into or going out of the KSC area. These weather calls are invaluable to the launch managers on launch days when weather becomes a factor. On the day of January 28, 1986, the weather aircraft had no weather calls to make. It was clear and cold, bitter cold!

The onboard Orbiter computers and telemetry processors continuously tell the ground computers and firing room engineers the health of the Orbiter, the boosters, the tank and all of the flight systems. Temperatures, pressures, valve positions, flow rates, critical launch commit decision data is transmitted to the ground via interconnecting data lines or by radio link at a rate of 128,000 bits per second. Ground sensors and data links advise the launch team of weather, ground systems status, and critical support readiness. Systematically, the data is decoded and processed by the ground processors! Two or three hundred systems engineers, all experts in their discipline, monitor these data and keep the Launch Director advised of the slightest indications of a problem or a negative trend that might be developing.

I remember that on the morning of January 28, we held the crew at their quarters until we had a good estimate for an expected launch time. The crew arrived at Pad B slightly after 8 a.m., and most of the seven were in their seats around 8:35 a.m.

During the beach house social three days before our first late January launch attempt, I joked with Dick Scobee, "Dick, I'm looking forward to you guys bringing Challenger back and landing at KSC."

After seeing Mike Smith go back for seconds on the barbecue, I warned him, "Mike, be careful on how much you eat. We have a weight limitation on this flight, you know!"

The very last words I spoke with Ellison Onizuka concerned a coke can. As I finished dinner, he grabbed my empty coke can and put it in the trash. "I'll do that," I said.

"No", Ellison replied. "I'll do cleanup, you go make sure that the Orbiter is ready to fly." I shall eternally remember this mild-mannered Islander who believed in the Shuttle team maybe more than any of the crew. He had followed closely the work we had done on the Approach and Landing Test Program. We had become admirers of each other's work.

I recall also how Greg Jarvis spent time that evening in serious consultation with one of his company engineers regarding some post-flight lab testing they would do on the hardware when Greg returned from the mission.

I remember Christa's parents and how they seemed to be in awe of the event and the astronauts. Christa's pleasant smile was as genuine as her obvious interest in going into space.

I recall that I never remember leaving the beach house after one of those barbecue dinners without thinking as I walked to my car, "This may be the last time I see these folks alive." And I always said a short prayer as I buckled my seat belt to leave, "Lord, protect and keep these great men and women safe from danger." I remember thinking the same thoughts about the Challenger seven.

These were the seven of Challenger as I knew them. They were Americana at its best, chosen to be the crew with which an American school teacher was to be the first civilian flown into space.

Each and every space mission is commemorated by a crew mission patch. Some of the most beautiful art is evident in these bright arrangements of ideas and symbols associated with the crew's mission. Some crews design their own patches. Some only suggest what they want and let designers do the work. Dick Scobee selected KSC engineer Ernie Reyes to design his crew's mission patch. Ernie, now retired, is one of the true KSC space experts, a multi-talented individual who can talk the horns off a proverbial "billy goat." Ernie can complete a complicated task in five minutes that would leave most engineers puzzled for an hour. He is a colorful product of a proud Hispanic heritage. Ernie and pro golfer Lee Trevino would have made a great "matched pair." He has won numerous honors and has been a contributor in key positions in all of NASA's manned programs. Ernie was the originator of the Snoopy astronaut award, based on his use of the Peanuts character to bring interest and pride to KSC and NASA's space programs and work schedules. Ernie's design of the STS-51L mission patch includes seven stars from an American flag, an Orbiter with golden plume zooming from earth, and of course, the names of the crew members. The one salient feature is the little red apple near McAuliffe, recognizing the school teacher on the crew.

It is also normal practice for the crews to send an autographed crew picture to certain people prior to launch. Prior to Challenger, I had always received my launch picture at least one week prior to launch. Although I never had the time to think of it, I did not receive my STS-51L Challenger crew picture before launch. It was late for some unknown reason. After the fateful accident, my secretary Barbara thoughtfully waited for some time before passing the picture along to me. Her thought was appropriate, because the picture read, "To the greatest Launch Director in the World," and it was signed personally by each member of the Challenger crew. No one alive today can possibly understand the deep feelings of sorrow that I felt when first reading those words. Today the picture hangs in my family room. I never look at it without being reminded of my dear friends, people I loved and admired greatly!

When Dick Scobee settled into the commander's seat in the Challenger cockpit, I called on the radio communications link to wish him a good morning and welcomed him and his crew to KSC's "northernmost launch site."

Dick humorously replied, "You mean if they'd launched us off the south pad, it would have been warmer, eh?" This was to be our first launch from Pad B which, in fact, is about one mile north of its sister pad, Pad A.

We had held the countdown several times during the night. We finally managed to get the special emergency work crews clear of the pad as we held at T-20 minutes. After T-20 minutes, things really started to happen at the pad and in support areas all over the world. During the hold at T-20 minutes, Arnie Aldrich again came to my console in the Firing Room and assured me that all concerns with the flight hardware had been addressed. He recommended that we proceed with launch. The contractors considered the ice to be of no significant debris concern and felt there were no constraints to launch. The "go-no go" poll at T-20 minutes included a "go" note from Rockwell International Corporation, the Orbiter design contractor; Martin, the external tank contractor; Rocketdyne, the main engine contractor; USBI and Thiokol, the solid rocket booster and solid rocket motor contractors. No one at any time, through any channel, attempted to halt the launch of STS-51L. There were procedures and policies in place to allow concerns to be surfaced! Neither I nor the Shuttle Program Manager Arnie Aldrich was informed of a joint seal concern by the Solid Rocket Booster community. Nor was Jesse Moore, the Associate Administrator for Space Flight, informed of a temperature limitation on joint seals. After we cleared the threatening icicles and refurbished the sound suppression system water bags, I was as confident about the safety of the crew as I could be.

I felt good about the launch. We decided to press for a launch time of 11:38 a.m.

The last nine minutes of the Shuttle countdown is almost entirely computer controlled. Computer talks to computer and as long as "go" flags are given, the countdown progresses. We were to listen later to the onboard tape recordings of the crew's banter over the intercom voice links as they commented on how cold it was that day. During the ten minute built-in hold at T-9 minutes, we again polled the same engineers, managers, and contractors to get a final "go-no go" for launch. Everyone associated with the Shuttle launch countdown understood the democratic process we provide by which anyone who does not support a "go" decision for launch can speak out loud and clear and be heard. We heard no objections to proceeding to launch Challenger!

During the last minute of the hold, the Launch Director says his last words to the crew. I wished my friend, Dick Scobee, and his crew a pleasant flight and promised to greet them all in a few days after they landed back at the KSC runway. The calm-voiced Commander thanked the launch team for its fine support.

On the roof above the LCC, Barbara, Bob Harris, and other astronaut escorts had led the astronaut families up to one of the prime launch viewing sites. Among the family members were Christa's husband, Steven, son Scott, 9, and daughter Caroline, 6 years old. They stood where I had stood on Sunday morning and observed the launch perfect weather conditions. It was breathtakingly cold in the clear air. Here the close family members would get a "bird's eye" view and be able to feel the sound and fury of this mighty spectacle. They could see the smoke billowing the last six seconds, as the three main engines came up to thrust. This was without exception the most beautiful day for observing a Shuttle launch to date. Within a radius of 100 miles, the white-tailed ascending spaceship could be seen winding its wings upward. Our KSC Associate Director, Al Parrish, often described the Shuttle ascent as a white-winged angel finding its way back to Heaven.

I was to be told later by Barbara and Bob that the Orbiter balloon that decorated my office the day before had somehow deflated. It was lying on the table when June Scobee and her family arrived that morning. Barbara recalls the comment June made at seeing the Orbiter balloon, "Gosh, folks, I sure hope this balloon is not a bad omen." I'm sure those were not June's precise words, but pretty much the impression she related to those present. We never know the true significance of such simple instances and surely most are merely coincidental. But, in the unfolding of an historic accident, I often ponder whether a power far beyond our intellect is providing insight along our pathway to prepare us for impending trouble. Does our God work that way? Again, only the revelation of eternity with Him will provide the answers to such remarkable questions!

Coming out of the last ten minute hold at T-9 minutes and counting, you sense the tension beginning to build more than ever in the Firing Room. A lot of us have learned to pray silently and reverently from T-9 minutes until the main engines shut down at about + 8 1/2 minutes. Everything is exact as you gaze from data screen to TV screen and back to data screen. At T-7 minutes and 30 seconds, the Orbiter White Room is retracted in order to clear the vehicle as it lifts off the pad. This swing arm is the crew's only route of escape from the Orbiter once they are inside and the hatch is closed. It can be returned back to the egress position in less than 30 seconds. The color TV camera in the White Room providing the means by which we watched the crew board the Challenger now gives us a scanning view of the Atlantic shore and the bright xenon lights and the surface surrounding the launch pad.

You learn to monitor valves open or close, tanks pressurize, swing arms retract, pyrotechnic initiators arm, auxiliary power units come up to speed, aero surfaces gimbal and position for launch. We were obviously exuberant in the Firing Room. Everyone loved this crew, these great men and women. I felt a comfortable feeling that all was right. We had identified our problems, and we had held until we corrected them. I could sense that 11:38 a.m. was going to be the right launch time. Nothing I have ever experienced in my career in the space program gives a greater thrill than to watch the indicators on your data screen turn "green" when they are supposed to. You soon learn to key your eyes and mind on two or three hundred significant parameters in the last nine minutes. You realize these functions must occur on time in proper operating order if you are to launch. The calm professional voice of the young woman operating the Ground Launch Sequencer calls out each major milestone as the clock proceeds to count backwards. Her voice displays the same coolness as the clock reaches T-31 seconds and she reports, "GLS is go for auto sequence start." This simply means that the ground software has sensed no out of limit conditions that would deter the Shuttle launch. It also says that the ground computer had relinquished control of the countdown to the onboard computers. In the perfect hold-free execution of a Shuttle launch by man and machine, I often felt as if I could feel what a symphony orchestra conductor feels as the musicians play the masterpiece. Music, a masterpiece of technical syncopation, magic! Those who have been involved with several pad aborts are also tuned to always expect the unexpected, the undesired! One must be ready to respond to abort situations at any second in the countdown. We have trained for such instances, and we have written the GLS software to aid in coming out of an abort situation with no adverse results.

At T-31 seconds, the onboard Orbiter general purpose computers take charge of the launch and control all the launch critical functions aboard the Shuttle. The onboard computers will continue to launch if they see no malfunctions and if the ground computer does not send a "hold" flag. Any hold sent after T-31 seconds results in a recycle to T-20 minutes, and in most instances, a scrub for the day. One old rocket scientist buddy of mine describes the last 31 seconds of a Shuttle launch as "the longest day of the week." Literally thousands of mechanical and electronic wonders occur between the flight and ground equipment from T-9 minutes until T-0. Your heart begins to pound and your eyes rush to take in a hundred last second images, data, people, TV screens, the room and atmosphere around you.

Around T-25 seconds, the ground launch software sends commands to open giant water valves at the pad. This action allows water from a 300,000 gallon water tank to deluge the flame pits underneath the Shuttle vehicle. This massive flow of water serves to provide sound suppression of the Orbiter main engine and help dissipate the horrendous heat generated at T-0 by the combination of main engines and solid rocket motors. This water, when impacted by the heat of the three main engine exhausts at T-6 minutes, creates the brilliant white plume of steam which bursts from beneath the Shuttle stack and propels a giant white cloud southward from the pad.

At T-10 seconds, giant "sparklers" are ignited on the pad surface near the Orbiter's three main engines. Simultaneously, you hear the familiar GLS call, "GLS is go for main engine start." These hydrogen burn igniters are designed to ignite and burn any free hydrogen associated with starting the three main engines. At T-6 seconds, the three main engines are started sequentially within fractions of seconds of each other. When all three engines reach the desired thrust level, the command can be sent to ignite the solid rocket boosters and release the huge hold-down bolts securing the Shuttle to the ground.

To most observers, the billowing white cloud of water vapor that is created by the main engine exhaust burning into the engine exhaust trenches lasts much longer than six seconds. The vapor is diverted southward and pushes away from the pad for two or three hundred yards. The thrust of the main engines and the fact that the Shuttle is restrained for six seconds results in pushing the tip of the External Tank in a northerly direction as much as four to six feet. Some of the astronaut crews report that it feels as if you are starting the first buck of a wild bronco ride when the Orbiter sways forward attached to the tank. The resulting reverse sway, referred to as "twang," rebounds the Shuttle stack back southward as much as 1 1/2 feet at the tip of the External Tank. Just as this rebound returns to dead center, when the stack is again vertical, the fire commands are sent to the solid rocket motors and the hold-down post pyrotechnic initiators. This is the long anticipated time of T-0. I always turned my chair slowly around to get a view of the launch from a location where only a handful of people have the opportunity to observe.

At T-0, the solids ignite, and the combined plumes tend to immerse the northern portion of the pad in brilliant white smoke with gold, pink, and red hues intermixed. The SRB plumes are "browner" than the pure white main engine exhaust plumes due to the flames produced by the burning solid propellant. The Shuttle rises slowly, and the feeling is totally exhilarating. At T + 6 seconds, the Orbiter computers command the aero surfaces to produce a 90 degree roll maneuver just as the Orbiter clears the launch tower. This points the Orbiter's vertical stabilizer, or tail, east toward the Atlantic. It also results in the crew being launched "on their backs" to orbit. This is a maneuver designed by the aerodynamics engineers to provide a maximum thrust profile for the ascent to orbit. The crew is trained to call out "Roll Maneuver" when they feel the vehicle turn. On one voice channel, I heard the JSC Public Affairs officer reporting the Shuttle's progress. On another, I heard the flight director team reporting ascent parameters and the calm voice of my buddy, Ascent Flight Director, Jay Greene, responding "Roger!"

What a magnificent view I enjoyed from my Launch Director seat in Firing Room Three that day. I pushed my chair back and watched proudly as the Shuttle rose through the deep clear blue Florida sky, trailing the long thin line of pure white behind it. The whole room vibrated, the windows rattled, the floor shook, and my chest pounded from both the dynamic pressure and the exhilaration. Tears often flow over the joy of a successful launch. Before your eyes, across the panorama of these massive windows, develops a perfect ascent trajectory! You always continue to say a silent prayer for the crew's safety. They are now riding this fiery chariot skyward! On this particular late January day, the Florida sky was so absolutely clear and heavenly blue, you could have seen a sparrow from ten miles away. I remember a feeling of peaceful comfort that we had managed to work through our adversities that beautiful morning. It felt so great to know we were beginning the mission that would bring telecasts to America's school children taught by a real All-American teacher. I was on top of the world, tired and physically weary, but never feeling more exuberant at another successful launch!

CHAPTER 21

"Go at Throttle Up"

...I bore you on eagles' wings and brought you to Myself.

Exodus 19:4

Go at throttle up! To those of us who have been closely associated with the Space Shuttle program, these four words have multiple meanings. When ascent of the mighty Shuttle assemblage reaches and successfully travels through Max Q (maximum dynamic pressure), we breathe a sigh of relief. For those of us who were close to the consequences and tragedies of the Challenger accident, "Go at throttle up" is yet another memory.

At +65 seconds into the flight of STS-51L, the Cap Comm (Capsule Communicator Dick Covey) at the Mission Control Center in Houston radioed to Dick Scobee, Challenger's commander, "Challenger, go at throttle up!" In his expected cool and confident voice, Dick Scobee replied, "Roger, go at throttle up." This was the last voice communication from the Challenger crew to the ground.

Throughout this 73 second flight, no one on the ground or onboard had observed any anomalous condition, no indication of any problem whatsoever. The Orbiter main engines had been throttled down to 94% thrust at +24 seconds. They were further throttled down by the onboard software to 65% thrust at +42 seconds. At +59 seconds, the engines were throttled up to full performance at 104% thrust. This engine throttling back in thrust puts the shuttle in the maximized aerodynamic condition to go through that layer of the atmosphere where the maximum dynamic pressure is experienced.

To the skilled flight dynamics officer, the Shuttle control systems reactions had been strictly to counteract the wind shears encountered during the flight. No one could possibly have seen and interpreted the sinister leak from the Solid Rocket Booster joint or the accompanying plume which would eventually destroy Challenger and its crew. The plume at about +64 seconds eventually torched its way through the thermal protection and skin of the External Tank. This extremely hot flame and the hyper-explosive hydrogen of the External Tank created a massive amount of dynamic, destructive energy.

As the Shuttle tank and strut linkage of the Solid Rocket Booster began to experience structural failure, hydrogen, oxygen, and the plume all combined to create a massive, destructive fireball. The Orbiter broke into several large sections when the tank exploded. The Solid Rocket Boosters continued to thrust in what appeared to be a guided fashion until destroyed at +110 seconds by the Range Safety Officer on the Cape Canaveral Air Force Station. At 78 seconds, the Orbiter's left wing, main engines, and forward fuselage (cabin) fell in a wild spiraling motion toward the clear blue Atlantic waters.

Within the next thirty minutes, the SRB recovery ships and an H-3 helicopter (Jolly 1) were heading toward the fallen Challenger debris. For many Americans, engineers, politicians, school teachers, pilots, family members, and the "man on the street," a few minutes of extreme tragedy had started a scenario that would redefine the way our country treated its heroes. The mighty chariot called Shuttle, once thought to be invincible, so easily accepted as routine by society, was now a vehicle for gloom to those who loved the adventure it represented.

America mourned with the families of the Challenger crew. Some found words to express the deep sympathy they felt.

We watched the Shuttle Challenger
Climb steady towards the skies,
Her bright orange flame of victory
Blazed sharply through our eyes...

Towards the threshold of our atmosphere
Did this mighty spaceship soar,
As the ground beneath us trembled
From her thunderous uproar.

Unexpected flashes overhead
Signaled trouble with the flight,
For seven lives we fearfully prayed
That they might be all right.

At heaven's gate their mission changed,
No further word from them,
For God had called them as His crew
And now they are with Him.

Tonight we share our common grief
And remember them with pride.
Let us carry on with courage --
The dream is still alive.

Written by Jim Ball, KSC Public Affairs
and read during a Memorial Service at
First Presbyterian Church, Titusville
January 29, 1986

President Reagan's memorial to the Challenger crew was:

The future is not free: the story of all human progress is one of struggle against all odds. We learned again that this America, which Abraham Lincoln called the last, best hope of man on Earth, was built on heroism and noble sacrifice. It was built by men and women like our seven star voyagers, who answered a call beyond duty, who gave more than was expected or required and who gave little thought of worldly reward.

Pres. Ronald Reagan, January 31, 1986

Excerpts from a speech by Senator John Glenn, the first American to orbit the earth, in a memorial service for Judy Resnick at Akron, Ohio:

Moments after Commander Dick Scobee spoke the crew's final words- "Go at throttle up- we watched, stunned as Challenger exploded in fire and smoke.

Now, even as we sort out the cause of the disaster, let us recognize those words not just as an epitaph but as a challenge – a solemn charge to continue the mission of the Shuttle crew.

The conquest of space is not merely a technological project of interest only to a handful of select scientists and specialists. It is nothing less than the expression of America's spirit.

In the brief moments of the Challenger's flight, we saw both triumph and tragedy of the human condition. We saw our own magnificence and our own mortality.

As we painfully learned, life on the edge allows not margin for error. But as President Kennedy once reminded us, we do not accept the challenge because it is easy, but because it is hard.

So as we reflect on Challenger's last voyage in the days and weeks ahead, I hope we never forget that the words "Go at throttle up" are far more than a courageous epitaph.

CHAPTER 22

The Aftermath

He breaks the bow and cuts the spear in two;
He burns the chariot in the fire.

Psalm 46:9

At T +73 seconds into main stage (SRM burn), all communications with Challenger ceased. The last word we heard from the Challenger commander was "Roger, go at throttle up!" This signified that the main engine thrust was back to normal following Max Q (maximum dynamic pressure). No more telemetry downlink! One big white cloud of smoke before us in the clear blue sky! Then I saw the two white rockets streak away from the main explosion. We all knew immediately that something terrible had happened. My first hope was to see the Challenger sail clear of the fireball and slowly circle to begin a controlled glide back to land safely at the KSC landing facility. Ace flier Dick Scobee, good old John Wayne, had taken manual control and would bring Challenger home. The alert crew had seen a problem developing that we had no visibility into on the ground. They had seen the danger and separated just before the explosion. These were the wishes of my heart, the things I prayed automatically would happen. Then the awful truth became evident to me! My technical mind overcame the wishes and desires of my heart. This concocted image of an intact Orbiter beginning its circular path to return safely to Kennedy never materialized. The crew cannot separate the Orbiter during first stage burn in ascent! It physically cannot be accomplished. The Shuttle design doesn't allow that option. The Orbiter had been destroyed. There were no parachutes, no abort rockets, no crew escape module, absolutely no hope! The five or six of us on the top row of the firing room pushed back in shock. No one uttered a word! One of our aerospace writers Chet Lunner at the KSC press site, on seeing the Challenger explode, said he first thought of a Return to Launch Site Abort. He ran to the KSC press building to tell a KSC Public Affairs Officer that the reporters were going to need a bus to rush out to the Shuttle landing facility. A photographer in the Mission Control Center in Houston caught Jay Greene, the Flight Director, with an expression of shock at the tragedy. This picture was to become one of the most publicized photos in the aftermath of the accident. There had been no warning to any ground controller of an impending danger. No malfunctions, no obvious failures, no negative trends! No data anomalies! The normally jubilant Firing Room team sat in silence for long minutes. Steve Nesbit was the Challenger mission commentator at the Johnson Space Center that day. A veteran "play-by-play" man for Shuttle launches, Steve's call at T +1 minute and 51 seconds was the most quoted by the media. With a calm professional voice, Steve reported, "Flight controllers are looking very carefully at the situation. Obviously, a major malfunction." I felt totally helpless as I watched debris falling into the ocean waters on the television monitors. This devastating sight was provided by the long range tracking cameras at the nearby Patrick Air Force Base. Helicopters were immediately dispatched to sea but warned to stay clear of the falling debris. Marv Jones, our Safety Director at the time, used the operational voice net and tried to bring some order to the first attempts to scramble boats and aircraft into the area. I remember hearing a radio net alerting rescue personnel to remain clear of the debris which was "falling all over the place." No matter what depth of training you have been through, no matter how professional you become, no team of individuals is ever prepared for the trauma of a situation such as the Challenger explosion.

I recall that I hung my head low almost between my legs. I so wished this could be one of those crazy dreams I had experienced so often and that I would awake and clear my mind of this disaster. But it was not a dream. It was real. This was history. I sat in shock and prayed to God that through a miracle the crew would be saved. I literally fought back the tears. My heart raced, and I wanted to gasp for breath. I almost felt nauseated as I thought of what horrific experiences these super people must have suffered. I think my first conscious thoughts were the feeling of wanting to pound my fist into my hand and cry out, "Lord, Lord, why did this have to happen? Why did seven wonderful courageous Americans have to die? Why did this happen when I was Launch Director? Why, Lord, is this thing happening? Why?"

During the last hour of the countdown, just prior to 11 a.m., my son Chuck, a ministerial student at Palm Beach Atlantic College, toyed with the idea of skipping the mandatory chapel service and watching the launch in the Student Union Lounge. He is a real space fanatic and especially enjoyed the thrill of the launch phase because his Dad was involved. At the last minute he decided to attend the chapel service. He was totally shocked when the chapel speaker was interrupted around 11:45 a.m. with a message that the Challenger had exploded during ascent, and the seven astronauts had been killed. As the students were asked to bow for prayer, Chuck said his first inclination was to pound his fist into his hand and say, "Lord, why? Why did this have to happen? Why did this have to happen to my dad?" Then he said that he felt the voice of God speak to him and say, "Son, don't you worry. Be still and listen. I am still in control. I am in control of the universe and all that will come of this."

It must have been at about the same moment in time when I found myself asking, "Why, Lord?" I felt the same peace come over me and the voice of God speak through my mind and spirit. "Son, you are not to worry, for I am still in control of this universe. I am in control of every situation and all will be right in this." I realized that this was the strong voice of God reassuring me of His power in creation and His love toward us in such ordeals. It is the peace of mind that only a true believer can know and understand. I was to understand later, reading from Psalm 46, the message of God to us in times like these: "Be still and know that I am God!" I felt the powerful promise, "God is our refuge and strength, a very present help in trouble."

We sat there on the top row of the Firing Room for at least thirty minutes. If I caught the face of someone in the group, I only saw dejected looks of unbelief. No one spoke a word! Center Director Dick Smith's eyes seemed to be trained out over the Atlantic. Even as thoughts came to mind, I found myself unable to put them into words. I'm sure all of the others who sat with me, Dick Smith, Shuttle boss Bob Sieck, Payloads Director John Conway, Marv Jones, Safety Director, and Challenger's Flow Director, Jim Harrington, all expressed these same types of feelings. They all felt the same inability to speak. At the end of our row of consoles, Hugh Harris, who had announced launch progress for KSC Public Affairs, was trying to find some constructive means of releasing information as best he could. We were all without any knowledge of what had caused the explosion. We could only speculate in our minds of the crew's condition. I heard the Range Safety representative reporting the location of the recovery forces trying to search the ocean's surface. I soon came to realize as my rational mind began to function again that the Challenger was destroyed and its crew was dead. Seven great men and women, my beloved friends and colleagues, were dead! None of the heroic rescue efforts seem to mean much to me. Then I realized that those who are rescue workers keep hope alive under the most pessimistic conditions. I learned to appreciate their zeal and determination to hold out for the chance that one of the seven could be alive and needing help. I felt deep in my heart that these efforts were futile. A national tragedy of historic proportions had just occurred. History, sad moments in history, had unfolded in the beautiful clear skies in front of us. I think that had I been a younger man, maybe in my thirties, I would not have recognized in real time that the nature of this day would live in the memory of so many people. January 28, 1986 would become another date in our nation's history that would be remembered forever.

Across the Kennedy Space Center and all over the country, wheels began to turn. Data was being reviewed. Television and movie films were being scanned and rescanned. Records were being impounded.

Log books, tapes, recordings, every imaginable form of information were being gathered to help determine the cause of the catastrophe. As in criminal investigations, any data pertinent to solving a crime must be protected. Our quality workforce is trained to immediately impound all official documentation, whatever form it is in, when an accident occurs. They did a top notch job that morning of January 28, 1986.

On the roof of the Launch Control Center and at the Public Affairs press site, the real personal suffering was being experienced by the friends and families of the Challenger crew members. At the press site, Christa McAuliffe's family cried with pain and utter shock. On the roof of the LCC, Bob Harris spoke to Barbara, my secretary, "We had better get these folks down. Something real bad has happened." The astronaut escort Frank Culbertson and the Public Affairs escorts handled the family crises in a highly commendable manner. There were later comments by some of the crew's family members that they were not informed of anything. They said they were not told of the crew's condition. I understand the shock and anguish these folks must have felt. But, in all honesty, no one knew much to tell anyone. I know we held no information from the media or the families that we felt was factual and would provide insight into the accident. NASA has never been a "cover-up" organization and, if anything, has been too quick to provide information. Most government agencies have such a "pecking order" chain of command that once information gets to the level where it can be released to the public, you hardly recognize what's being discussed. NASA does not operate that way. Everyone has the opportunity to speak their mind without reprisal. So it was in January 1986. We tried to keep everyone informed as we learned the facts. We kept someone close to each crew member's immediate family and moved the families to a private place as quickly as possible. We gave them as much credible information as could be determined. NASA hastened to assign one of the medical personnel to each of the astronaut families. I realize today that no matter how low we felt on the launch team, we could not know the suffering that the crew's families were experiencing.

After about thirty minutes of sitting solemnly in the Firing Room, one of the armed guards who was assigned as a monitor for Firing

Room access approached the top row where we sat. He informed me that there was data in one of the television recording rooms downstairs that someone had requested we come down and review. I remember distinctly the key engineering managers forming a single line as if we had practiced it for days. Slowly, almost ceremoniously, twenty or thirty of the top space scientists and engineers in the entire world marched out of the Firing Room and boarded the two elevators waiting for us. We were escorted to a recording room in the northwest corner of the LCC's first floor. I don't recall that a single word was exchanged between any of the people in the group. Only total silence!

In the record and playback room were hundreds of data and video recorders that preserved every TV camera scene, every voice net, every word spoken between launch team members, and every data stream flowing to and from the Shuttle. I was pleased to see that some of our veteran engineers were still dedicated data gatherers. I had always appreciated the advantages of having good quality records. I never appreciated it more than in the investigative phase of the Challenger reviews. We all stood around anxiously hoping to discover some amazing revelation to the cause of the accident. We were repeatedly shown different video accounts of the Challenger launch, ascent, and the violent explosion. After two or three viewings of the videos, I found watching the explosion over and over to be quite an emotional ordeal. At times, several people commented about the bright glow seen near the external tank attach points prior to the disaster. We all knew, even that early in the investigation, that something off nominal had caused the bright spot. Perhaps there had been some kind of leak or flame either emanating from one of the solid rocket motors or from the external tank. It was impossible to determine the source of the leak from the film we reviewed that day. Later on, this type of documentation was to be the most valuable data used in pinpointing the Solid Rocket Motor joint leak. We probably watched the dreadful replay a dozen times at various speeds trying to glean some valuable bit of information. Each rerun I watched only added to my personal feelings of grief and sorrow for the men and women of Challenger.

After no concrete results from viewing the videos, the management team began to recover from the shock of the disaster. It was almost as if we were "coming to our senses." I directed one of our best operations engineers, Chuck Henschel, to secure a special conference room area on the fourth floor of the LCC. We all gathered in this specially designated area to begin to organize for the many investigative jobs ahead of us. We left a few booster and tank experts to continue studying the videos for clues to the cause of the accident. An armed guard was posted at the entrance to our management area, an access list generated, extra phones and desktop computers brought in, and most of NASA's senior managers were assembled. We initiated a "war-room" type operation in the conference room and began to delineate action items. One of the very first actions was assigned to me. I was asked to designate someone from the Kennedy Space Center to lead a Challenger debris recovery effort. I walked out into the hallway of the area that was assigned as offices for the people in my organization. I looked down the hallway, and the first person I saw who reported to me and worked in my organization was a young engineer, Elliot Kicklighter. With a lot of haste and little other thought, I decided that Elliot would do an excellent job based upon my knowledge of his previous job performances. I remember the startled expression when I grabbed his arm and said, "Elliott, I want you to be in charge of the debris recovery operations." I am a firm believer in making decisions when you need to. I don't relate well to people who would rather pass the buck than make a decision. My idea of a good manager is that he or she basically knows how to make a decision. My decision to select Elliot turned out to be a very good one. He did an outstanding job in an assignment that required long stressful hours over a long period of time. I know it was the bountiful grace of God that led us, all tired and dejected, through those first few hours of crucial decision making in the aftermath of the tragedy.

Arnie Aldrich and his able deputy, Dick Kohrs, led the initial evaluation of the action items that needed to be worked. Dick was by far the best integrator NASA has produced. His sharp mind and keen insight were invaluable to the management team in those first efforts to recover from the shock. As usual, Dick worked the details by actually serving as the scribe who listed the action items on the chalkboard.

One of the first questions asked by someone in the group was, "Does anybody have a contingency plan?" Dr. Bill Lucas, the erudite Center Director of the Marshall Space Flight Center, opened his neatly arranged briefcase and there in the cover portion was a green-faced document entitled "STS Contingency Plan for the Marshall Space Flight Center." I immediately grabbed a nearby phone to muster up a copy of the KSC Contingency Plan, if such a document existed. I am embarrassed to admit that the Shuttle Launch Director did not carry a copy of this important document to the Firing Room for launch. Neither was he familiar with its contents. We were almost naive in our neglect of considering contingency planning seriously. We had planned well for aborts and other emergency situations. We never really seemed to foresee a catastrophe of such tragic magnitude. I'm afraid we may have lulled ourselves into thinking we were impervious to tragedy and failed to put the proper emphasis on reactionary practices. I would like to believe that someone at KSC knew the contingency plan from memory. Surely someone was the office of primary responsibility for such a crucial document. No such person or persons ever surfaced. One of my primary actions after Challenger was to establish a Contingency Planning Team and make this process one of our highest priorities at KSC. It has become a well-established function and is reviewed and updated on a regular basis. We finally located our KSC Contingency Plan on the fourth floor, Center Director's staff, of the KSC Headquarters Building. I swore never again to downplay the significance of contingency planning. We would forever have and maintain contingency procedures for all credible failures or accidents. Again as I remembered my inadequate handling of the hatch tool fiasco, I noted that even though you feel you have fared well under fire, you may have suffered far more wounds than you felt on the day of the battle.

One significant action after another was assigned. Data impoundment was emphasized and a responsible person or organization assigned. Recovery operations! I took care of that early by assigning Kicklighter to head up this team. Family welfare fell to the Astronaut Office, the Medical Officers, and Public Affairs. All records, data recordings, tapes, log books, any and all forms of possible information were collected. All these data were impounded and locked under constant surveillance. My own personal logbook, an ever present companion, was confiscated. I was always a stickler for keeping a log to serve as a repository for significant notes or ideas. We also used it to communicate and to keep informal accounts of our activities. I have no idea of what kind of notes I made during the course of the Challenger countdown since I never saw my logbook again. I inquired on several occasions as to anyone's knowledge of where personal logs were impounded. I never located it. I assume that if it had held any enlightening words pertinent to the accident, they would have been part of the Rogers Commission's investigation reports. It also became clear to me that "personal log" is a misnomer. There are no such things as "personal logs" in the aftermath of a major incident such as Challenger. I would not want it to be otherwise. I consider what we record concerning the conduct of our job responsibilities to be part of the official functioning of our work, especially so for the federally employed.

I began to notice astronauts come into the conference room, a real tribute to the manner in which the men and women of the corps rallied behind us and offered to help us in any way they could. Although I never asked him to confirm it, I'm sure George Abbey, the Chief of Flight Crew Operations at the time, was instrumental in getting the many astronauts in their T-38 jets and on the way to Florida. We welcomed their support and they were to play vital roles in every phase of the Challenger investigations. Sometime around 2 p.m. on the afternoon following the Challenger accident, someone thought of food for those of us on the management team. Someone brought in several large boxes of Kentucky Fried Chicken and all the trimmings. I was told later that it had been donated by Kentucky Fried Chicken, because we were flying the chicken egg experiment on board. I don't remember anyone ever opening a box of it! I know I could not have eaten under such circumstances. I had not eaten a bite since dinner at home in the late afternoon of the previous day. Somehow, no one in the room had any interest in food. I'm quite certain that we never opened a single box of the chicken. It was not a time to eat!

We worked on into the early evening hours, planning what needed to be done, both short-term and long-term. I don't remember any reference to time. To me, time was standing still. Although my mind was clear as far as details and what we were accomplishing, my psyche seemed almost to be in a state of suspension. I'm sure the lack of sleep and nourishment were overcome by the flow of adrenalin. It is so rewarding to see everyone of one mind and accord when technical personnel convene under dire circumstances such as these. The actions which were discussed were obvious to all the people involved as being necessary. No one tried to shirk his responsibility and gladly accepted each assignment. Often one manager would volunteer support to the action assigned to another manager if he felt he could contribute. Very quickly six to eight hours elapsed as we began to form our plan of action. Most of us who had been around for a few years knew full well that we would get a lot of help from outside the Agency. And we certainly did! The first actions we defined and kicked off that day were crucial ones. But by no means were they all-encompassing. They were just a handful of the thousands of actions that would originate in Congress, in the Administration, and mostly from the efforts of the Rogers Commission appointed by President Ronald Reagan.

We received notice late in the afternoon that Vice President George Bush was flying from Washington to the KSC to meet with the crew's families. He also requested to speak to the KSC Launch Team from the Firing Room. Once again at 8 p.m., the Launch Team gathered in Firing Room 3 from which Challenger had been launched that fateful morning. I stood a few feet from Vice President Bush as he spoke. Senators Jake Garn and John Glenn stood by his side. I was deeply impressed by Mr. Bush's attitude and sincerity as he spoke to us of the need to get the investigation started, to find our problems, to fix them, and to press on. At times, when he spoke of Challenger's crew and its families, he cried. I will always remember how this great leader actually shared in our sorrow by coming close to us where we worked. He reassured us that we had the full support of President Reagan and the Administration. He told us we would find the cause of the accident, we would correct it, and we would carry on the space program. I shall always appreciate the empathy shown by this great Vice President and the tears in his eyes as he concluded his brief address to us. As I glanced around the Firing Room, I saw a very high percentage of the hundreds of engineers and console operators who had been at work since we first began tanking the evening before. I don't believe a single member of the launch team went home that day.

At very nearly the time we were meeting with Vice President Bush, President Reagan was addressing the nation on television concerning the Challenger accident. Remembered as a great communicator, President Reagan's words were most encouraging in this time of national mourning. Some of the things he said included:

I've always had great faith in and respect for our space program and what happened today does nothing to diminish it. We don't hide our space program. We don't keep secrets and cover things up. We do it all up front and in public. That's the way freedom is, and we wouldn't change it for a minute...

We'll continue our quest in space. There will be more shuttle flights and more shuttle crews and yes, more volunteers, more civilians, more teachers in space. Nothing ends here; our hopes and our journeys continue. I want to add that I wish I could talk to every man and woman who worked on this mission and tell them: "Your dedication and professionalism have moved and impressed us for decades. And we know your anguish. We share it...

The crew of the space shuttle Challenger honored us by the manner in which they lived their lives. We will never forget them, nor the last time we saw them, this morning, as they prepared for their journey and waved goodbye and "slipped the surly bonds of earth to touch the face of God."

CHAPTER 23

A Time to Weep

To everything there is a season,
A time for every purpose under heaven…
A time to weep…

Ecclesiastes 3: 1, 4

Sometime around 10 p.m. on the night of January 28, 1986, I finally felt the need to go home. I had been on duty for nearly 24 hours. So much had occurred in those 24 hours. Only the wildest of dreams could have packed so much into one day. Thank God, the adrenalin had kept me awake, alert, and sane. Many of us were almost to the point of breaking down from physical and mental exhaustion. Add deep sorrow to the exhaustion and you provide stress that can bring even the strongest person to their knees. I don't know how I managed to drive home. I drove in silence down Merritt Island, completely unaware of the surroundings. I pulled onto the driveway, turned off the car's engine and lights, and just sat. It had to be close to 11 p.m. But time still seemed to hold no relevance. I saw my beautiful sweet wife and two precious daughters open the kitchen door where they had been waiting. I still couldn't manage to move. My wife opened the door and asked, "Can you come on in?" I put one arm around her and we walked slowly into the house. No one said a word. We went on through the house to the master bedroom and the three of them sat close to me on the bed. Immediately, I began to cry like I had never cried before. I cried out loudly as if the family could provide answers. "Why in the world did this have to happen? I just don't understand why this had to happen. My good friends are dead. Oh, Lord, why, why?" My body shook as I wrung my hands and cried. I held to the ones I love so dearly. They knew they needed to let me cry and just hold on to me. I recall that first Wendy and then Karen began to cry along with me. I heard some people say that men should not cry, but I was never ashamed for that moment. My whole body cried out for understanding. I know today that those twenty minutes or so of crying was the best therapy I could have experienced. I felt no weakness as a man who needed to cry. I am convinced that truly there is a time to weep as the Bible so beautifully instructs. My weeping was not for me. I did not weep for NASA or the space program. I did not weep for the beloved United States of America. I wept for the seven individuals that I loved. I did not want to accept the tragedy of their death. I realized in my grief that I would be affected by this ordeal for the rest of my life. An indelible mark had been made in my personal life and in the lives of many millions of people who were mourning in their own way for these seven astronauts.

Somehow I managed to fall asleep late that night and slept in total exhaustion. I had no dreams that night, no tossing or turning. The next morning when the alarm woke me, my first conscious thoughts were to wonder if I had again experienced one of my horribly vivid Shuttle nightmares. All too quickly I realized that the day before had been no dream. The Challenger disaster had been real! I drove to the Center anticipating a day of hard work attempting to get to the answers of how and why the Shuttle failure had occurred. On our regular 8 a.m. "white phone call," I remember giving a short message concerning the tragedy we had all experienced the day before. It was a very rare occurrence when I spoke on this early morning "working level" teleconference. As a leader, I felt responsible to try to encourage the work force as best I could. In reality, I could only speculate as to how much of an impact this disaster would have upon our nation's space agenda. So I spoke from the heart without trying to make promises that I could not fulfill. I spoke of the tragedy that had happened, the desire to work as a team to get to the solutions, and the need to press on to keep the Shuttle program alive.

Our Center Director Dick Smith addressed the launch processing contractors and NASA Shuttle personnel on a special hook-up for the white phone call. Dick, with his special ingrained credibility, acknowledged the grief shared by everyone on the Shuttle team. He reminded everyone that NASA had a history of facing up to its problems and growing better. "What has made us great is that we have been forthright and honest in facing our problems and solving them." He assured the people of the KSC that President Reagan intended to continue to support the space program. He encouraged everyone to cooperate fully with the investigations that were to follow. That was to be our number one priority. In the meanwhile, we were to look hard at all of our procedures and how we do our job. He concluded by stating "we will be better in the future."

Somehow I knew in my heart that what I had said and the short speech by our Center Director were merely temporary words of encouragement. The Shuttle program would, and did, undergo months of intense scrutiny. Our people, our processes, and our procedures were taken apart piece by piece. One review followed another! It was not unusual to be appearing before or being interviewed by three or four separate review panels in a single day. I was tasked to chair a review of every major integrated test procedure used to process the Shuttle. Our two afternoons per week meetings were attended by 50 to 60 people. They lasted four to six hours, long and tedious skull sessions. We wrote and rewrote hundreds of procedures in order to assure no mistakes could possibly be made. It was a labor intensive job. Some critics think we over-reacted and made too many changes. If it would preclude another Challenger, no amount of time and effort would be too much.

I worked the day after the Challenger accident in an almost dazed state of mind. The initial shock was over, but the bad memories were beginning to plague most of my thoughts. For our family, January 29 would normally be a fun day for it is my daughter Karen's birthday. Not so on January 29, 1986. When I returned home that evening, I didn't feel like dinner. I chose to sit by the backyard pool in the cool evening air. I just wanted to think. I had meant to pray, but I'm not sure if I did. Just as I had tried to relax, I heard a car drive onto our driveway in the front of our house. My wife's voice greeted the two men who came to the kitchen door. I later learned that the two were newspaper reporters from the Orlando *Sentinel.* They were asking to interview the Launch Director about the Challenger accident. I could hear distinctly in the clear January night as Juanita stood with them on the driveway and evaded their questions. She deftly avoided telling them where I was. We later laughed as we recalled how she "reversed the charges" by asking each of them a lot of questions before they could question her. "Where are you guys from?" she asked one of the reporters. "And how long have you worked for the *Sentinel*?" she inquired of the other. "How many children do you have?" She must have "interviewed" them for close to an hour before they became discouraged and left.

At work the phone rang regularly with a request to speak to either a reporter or a television news person. I was surprised how many common people called to ask questions, give you the cause of the accident, and how to fix it, or to offer their condolences and support. Some of our people who were involved in the launch of Challenger became extremely angry at the dogged pursuit of the media. One manager's wife called the local newspaper to complain about the same two reporters who my wife had managed to get rid of. This lady's complaints were valid. They were just directed to the wrong newspaper! I tried not to react negatively to the media because I knew we needed their support. They had always treated me fairly in my tenure as Launch Director. I tried to understand their search for information. But I also needed to spend the first several days after the accident with some degree of privacy when I was away from the job. I intended to cooperate fully with the media after I could get peace of mind in my own personal feelings.

A lot of good friends began to stop by the office at work. Some sent cards and notes, small ways of expressing concern over the situation. I shall forever cherish the kind words and thoughts from so many friends and co-workers. The true character of a lot of fine people became clear to me. In so many cases, the "insignificant" people, the working level folks, were the first to pledge their support.

I understand better now a quote by hard-nosed newsman Walter Winchell, "A friend walks in when the whole world walks out."

On the Sunday following the accident on Tuesday, my family continued our regular routine of Sunday morning worship at First Baptist Church on Merritt Island. When I stood with our congregation to sing the first hymn, I knew I would have trouble singing with the normal joy that a church service brings. The hymn chosen was *It is Well with my Soul,*" its inspirational words written by Horatio G. Spafford. I almost choked with sorrow as those around me began to sing these peaceful phrases. I could not sing a single word, and my eyes were flooded with tears. I considered slipping out, thinking I might begin sobbing aloud. The first and last verses of this great hymn contain these words:

When peace like a river, attendeth my way,
When sorrows like sea billows roll,
Whatever my lot, Thou has taught me to say,
It is well, It is well with my Soul.

And Lord, haste the day when my faith shall be sight,
The clouds be rolled back as a scroll,
The Trump shall resound and the Lord shall descend,
Even so it is well with my soul.

The words about clouds, sea billows, and the sky rolled back as a scroll were all too associated with the Challenger for me to utter. As giant tears flowed down my face, people who had been watching my reaction began to pass their handkerchiefs or tissues along to me. Some of them also began to cry. I felt the kindness of their Christian love and knew that hundreds of friends were praying for those close to the Challenger. Although a very poor excuse for a singer, I had always enjoyed participating in congregational singing in church. Now these words only brought tears of sorrow, and a lot of handkerchiefs. I must have carried a dozen handkerchiefs home after the service. I told someone I could have easily opened a haberdashery and specialized in men's handkerchiefs. It was a sad experience to not be able to participate in the singing without crying. But attendance at the church service was good for the spirit and I realized that by the grace of God we are able to withstand the black days we must all go through. For several years, I was not able to sing a hymn in church without the memory of Challenger coming back. So I elected to listen while others sang. With the passing of time, I have been able to sing again occasionally, but never without the memories. I know it was probably attributable to our music director's repetitive scheduling, but it seemed that every January around the time of the Challenger anniversary, we always sang *It is Well with my Soul.* Each time the memories came back along with the tears. Again, it might have been the handiwork of God's touch inspiring the selection of this hymn to be sung. Some might consider the crying to be sentimental overreaction, but it was not.

For everything there is a season, A time to weep…

In my time of spiritual need, our pastor at the time, Len Turner supplied the wonderful encouragement that our family needed. Rev. Turner was a strong supporter of the men and women who launched the Challenger. His private prayers and words of support and reassurance from the pulpit have endeared our families for eternity.

CHAPTER 24

Recovering and Changing

From the ashes of disaster,
Grow the roses of success.

From a song in the movie,
Chitty Chitty Bang Bang

There were a lot of news stories in the early months of 1986. Corazon (Cory) Aquino, age 50, was sworn in as the new president of the Philippines after Ferdinand Marcos went into exile after long and bloody efforts to oust him finally succeeded. Meryl Streep was voted the all-around favorite female entertainer and Bill Cosby the favorite male entertainer.

On Wednesday, January 29, the day after the Challenger, the New York *Times* devoted its entire front page and nine other pages to coverage of the disaster. The last time this prestigious paper had devoted the cover page to a single event was the New York power blackout in 1977. Other events receiving total front page coverage included the bombing of Pearl Harbor, President John F. Kennedy's assassination, and President Nixon's resignation.

The days after the Challenger accident were hard. Not a single day went by that in my private thoughts my mind went through the series of events leading up to the accident. I awoke almost daily to first think I had been experiencing a nightmare and Challenger had been a dream. There had always been the sordid sleep images in mind of horrendous Shuttle disasters from which I awoke shaking, sweating, or shouting. I suppose a lot of people who are dreamers experience bad dreams when they work in hazardous work situations. But all these dreams were never without some crazy situation that could not be real. In my dreams the Orbiter would often land safely in my backyard. Bad dreams! There were never any credible scenarios, but a mixture of mental images forming a kaleidoscope of fright to the sleeping mind. The Challenger accident was not to be such a dream! It was real! Its scenario was understandable, though never expected.

If there are such things as encouraging words in the aftermath of a devastating accident, Congressman Bill Nelson's appearance on national television on Wednesday after the accident sure made a lot of us feel a little better. Representative Nelson, appearing on the *CBS Morning News,* was asked by Maria Shriver whether the Challenger disaster showed that NASA was compromising safety.

"No, ma'am, that's just simply not true," Bill responded. "As you know, we had seven delays on my mission on Columbia. NASA was not about to attempt that launch until everything was right and so too with Challenger yesterday."

Nelson pointed out further that NASA had done everything possible to insure Challenger was launched safely, but that a mistake had been made. "We've got to find what it is and correct it."

Also on Wednesday, February 5, 1986, the families of the Challenger Mission 51-L crew released this joint statement:

The 51-L crew families want to thank the people of our country and all the countries of the world for their thoughts, their feelings, and words of encouragement.

Space flight serves as an outlet for our human need to learn and expand. What's out there will make our lives better on Earth and help satisfy mankind's natural curiosity to explore and push the borders of the "known universe."

So that their lives were not lost in vain, we must rededicate ourselves to the exploration of space and to keep the dream alive.

In the first week or so following Challenger, there was a lot of speculation regarding the fate of the crewmen after the violent explosion. An outstanding job by Navy debris recovery forces covered thousands of square miles of ocean off Cape Canaveral. The National Safety Transportation Board sent its finest investigators to help us identify and piece together the thousands of Shuttle parts that were reclaimed from the ocean's floor. Debris was sighted and recovered all along the eastern shore of the United States from the Carolinas to the Florida Keys.

Sensation writers and tabloids had a field day at any recovery associated with the Challenger's crew. On a Sunday, March 9, 1986, remains of the Orbiter crew cabin were found in about one hundred feet of water about eighteen miles off Cape Canaveral. On Wednesday night, March 12, the salvage ship USS Preserver returned to Port Canaveral with a flag-draped coffin on its deck. There arose an unfortunate controversy, fanned by the press, over whether NASA's medical personnel and pathologists from the Armed Forces Institute of Pathologists should perform autopsies on the crew members. Others felt strongly that the Brevard County, Florida coroner's office had legal jurisdiction over the remains and should perform the postmortems. NASA was accused of mishandling the crew remains in an attempt to treat them as a special case and avoid local forensic involvement. I have no doubt that NASA was acting in the best interest of the astronaut families in trying to protect them from unnecessary disclosure of information regarding the crew's remains.

I strongly believe that there should be limits to the information given to the media. Sordid details of murders, accidents, or abuse need not be disclosed just because the public thinks it has a "right to know." I have become convinced that the media will expose practically anything to make a headline, scoop a competitor, sell a newspaper or periodical, or to make the television newscasts more exciting at the end of the day.

I am not intending to be harsh on the media in any way. I dearly loved and respected the dozens of regulars who covered our Shuttle launches. They quizzed me with respect and were never cruel in their search for information. But the Challenger accident, as could be expected, brought media attention from thousands who couldn't spell Orbiter if they were asked to.

For reasons of not needing to know and truthfully not wanting to know, I purposely tried to avoid any knowledge of the crew remains. I was asked repeatedly for comments! Thank God, I did not have to give answers. We were all fairly certain that the crew cabin survived the initial shock of the explosion for a few seconds. Some of the recovered hardware and post flight data indicates that Ellison Onizuka may have been able to turn on Mike Smith's personal emergency air pack (PEAP). Voice tapes revealed that the last voice recorded was "Uh, oh," probably spoken by Challenger's pilot Mike Smith over the intercom voice circuits. The cabin fell for about 2 1/2 minutes and impacted the water at near 200 miles per hour. Anyone who would provide an accounting of the crew's actions, what they experienced, and the accompanying trauma can only be speculating. I choose not to build a death scenario, but rather to remember them as I knew them: a vibrant, living group of people I dearly loved.

The Challenger memorial services at the Johnson Space Center held on January 31 were magnificent. They were a solemn, but fitting tribute to America's space heroes who gave their lives in pursuit of discovery. A group from KSC's NASA leadership was invited to represent Kennedy and its employees. Dick and Louise Smith, Tom and Judy Utsman, Bob and Nancy Sieck, Jim and Jean Harrington, and my wife Juanita and I flew to Ellington Field near JSC aboard our NASA 4 twin-engined Grumman airplane. There was no gaiety or relaxed atmosphere that day as was the normal experience when flying the NASA plane. We arrived after a nice smooth flight to find we had to sit for nearly an hour on the Ellington runway as Air Force One had just recently landed. President Ronald Reagan and Nancy led a top level government party to pay respects to the Challenger families.

We were ushered via government vehicle to the plush campus of the Johnson Space Center. George Abbey, the astronaut's boss, met us at the lobby of Building One with a hug. I had never before seen this man of action so soft and caring. I had come to know George closely from a distance. A strange way to describe a relationship, but true in all respects! We were friends because our jobs drew us together as comrades, fellow laborers in a grand enterprise! I recall as we looked at each other, we hugged as two men who appreciated and understood what the other was going through in his mind and in his heart. I remember as he invited us to follow him to the site of the ceremony that big tears flowed down the face of this man who rarely showed emotion. Like the experience with Vice President Bush, I again gained a new deep respect for another man who cared enough to let his tears be seen! To this day, George Abbey, later to become the Director of the Johnson Space Center, remains a friend I respect and admire. The families and loved ones of all of the Challenger's crew were at the ceremony. Most of the astronaut corps was present. Congressmen and diplomats were numerous. Because the President was there, security was heavy and agents were abundant. The program included a trumpet solo by Guy Bluford, the invocation by Astronaut Charlie Bolden, and prayer by Steve Hawley's father, Reverend Bernard Hawley, an ordained Presbyterian minister. President Reagan spoke eloquently of the Challenger crew, spending a few moments reflecting on the life and character of each of the seven crew members. What a touching tribute from the nation's top official! Years later, Edmund Morris, Reagan's authorized biographer was to call Reagan a great communicator, "He could articulate significant world events such as the Shuttle Challenger accident." Concluding with the missing man jet formation flying directly overhead, the ceremony was hard to sit through without breaking down in tears. Even those who were not as close to Challenger were tearful as the tragedy of such a blow to the American pride was memorialized. President Reagan took several moments to pass from one family member to another until he had expressed to each of them the deep regrets that I know every American would have had him speak. There is no possible way to find the words to describe my utter feelings of sorrow as I sat through the ceremony. Our language does not suffice when your deepest feelings are evoked. Inside you feel totally incapable of keeping your wits together. Thanks to God that He provides a way and we are never put to a test or trial through which He is not able to sustain those of us who belong to Him. I also felt the need to be near my sweet wife in such a time of mourning. Dick Smith was an astute leader to direct that we take our wives along. He surely knew that when you mourn the death of loved ones, you need those closest to your heart at your side.

The management of the Kennedy Space Center held a brief memorial ceremony for the crew at the External Tank basin. Kind words of solace were directed to the thousands of KSC employees who gathered on February 1 to memorialize the crew. Dick Smith and Astronaut Bob Overmeyer spoke briefly and three local pastors expressed the feelings that all of us were experiencing. I found myself again shedding tears of sorrow over the loss of seven wonderful people. I found it especially difficult to look toward the launch processing buildings and the two Shuttle launch pads where we had worked side by side with these crews. I was glad to see the ceremony short and succinct.

As a closing part of the ceremony, at exactly the time of the Challenger disaster, we carried out a special tribute to the memory of the seven fallen astronauts. Our Center Director, Mr. Smith, had called me a few days earlier and asked that either I or someone I designated close the KSC Challenger memorial service by dropping a wreath over the ocean offshore from the Center. The wreath would be in memory of the Challenger crew. I felt that I needed no more remembrances of the crew to deal with. I by no means wanted to load someone else with what could be considered a burden. No one was more closely involved with Challenger than its Flow Director, Jim Harrington. After a good deal of consideration, I decided to offer Jim the opportunity to drop the wreath at sea. Jim graciously accepted the honor and afterwards I knew we had made the proper choice. I shall forever remember, as *Taps* was played, Jim's bowed head and attitude of reverence as he carried the large garland wreath and boarded the NASA helicopter. The large red, white, and blue wreath contained a red blossom for each of the seven Challenger astronauts. As expected, the media covered every aspect of the activities. As if part of the script, several dolphins surfaced and swam around the spot where Jim dropped the wreath about three miles offshore in the Atlantic waters. Several newspapers reported that seven dolphins greeted the memorial wreath. Jim recalls that there were at least a half dozen dolphins repeatedly breaking the ocean's surface, but he could not get a count as to how many there actually were. Nonetheless, the presence of the sea mammals added an even more mystic tone to the memory of the day.

Two months later on April 29 seven black hearses carried the seven crewmen's remains from the Life Sciences Building on Cape Canaveral Air Force Station through the Industrial Area of the Kennedy Space Center to the Shuttle Landing Facility. From this extra long concrete runway built along the flat land north of the giant VAB, Dick Scobee and Mike Smith had practiced dozens of Orbiter landings in the Shuttle Training Aircraft. Hundreds of Kennedy employees lined the roadways and bowed in silent prayer as the crew's bodies passed by. I chose to remain quietly at my desk and watch coverage of the procession from the Kennedy television network. Officials from all over NASA formed a line leading from the hearses to the rear entry ramp of the huge Military Airlift Command C-141 aircraft.

In the aftermath of the historic accident and despite an enormous amount of investigations, primarily the Rogers Commission investigation, NASA began to change leadership. Dr. James Fletcher was asked to return as NASA Administrator and was given the job of "repairing the USA's troubled space agency." Dr. Fletcher, 66, had been NASA's Administrator from 1971 until 1977 and was described as "technically inclined, plus he has good managerial talents." Every comment I had ever heard about the distinguished gentleman was excellent, and he was certainly the best man for the imposing task of leading our space program back to prominence.

Another excellent leader Rear Admiral Richard (Dick) Truly, 48, was brought back to NASA to head up the Manned Spaceflight program and specifically to lead the Shuttle's return to flight activities. Truly started his spaceflight career as one of the original pilots chosen to fly into space in the Air Force's Manned Orbiting Laboratory program. When the program was canceled, Dick had transferred to NASA's astronaut corps. A veteran of the Shuttle program, Truly last commanded STS-8, a Challenger flight in 1983. My wife and I had been high school mates with Dick Truly at Meridian, Mississippi in the early 50's. Juanita and Dick were recreational instructors together in a program that provided top notch high school students as mentors for the city's elementary schools. Our friendship with

Astronaut Truly continued when we were on temporary duty at Edwards Air Force Base in 1976-77. Then Captain Truly was one of the NASA's astronauts who flew the Approach and Landing Tests aboard NASA's first Orbiter, Enterprise.

Richard was head of the Naval Space Command when asked to return to NASA. An exceptional leader, a gentleman, and friend to his co-workers, he led the manned space program through one of its darkest hours. Truly, in turn, selected some of NASA's best as managers, including a good number of astronauts. As the Associate Administrator for Space Flight, Truly outlined a comprehensive strategy and assigned major actions to return the Shuttle to flight with strong emphasis on safety. The recovery plan included a total Solid Rocket Motor joint redesign, a bottoms-up requirements review, a complete review of all critical procedures, guidance as to the first return-to-flight mission, and the development of a sustainable, safe flight rate for the Shuttle fleet. Richard's dedicated example and leadership earned the respect of all of NASA and its employees. He was later to be appointed as NASA Administrator, replacing Dr. Fletcher.

As NASA began its recovery process, we worked tirelessly at the Kennedy Space Center to re-engineer every facet of our operations. We received a lot of help, most of it certainly welcomed as constructive in rebuilding our procedures, our management structure, and our renewed dedication to safety, quality, and reliability.

As I expected, we also received a lot of help from the media and from special interest groups outside NASA. By far the most annoying assistance came from a group I shall refer to as "nuts." It's common to get a lot of false confessions and tips when a major crime is perpetrated. A lot of amateur detectives always want to help law enforcement officers solve a mystery. So it was with Challenger's investigations. A lot of well-meaning "experts" from all walks of life wanted to help find the problems and provide solutions to them. Very few of them had any degree of credibility!

One particular engineer was convinced that NASA had somehow damaged the Shuttle stack when we made the "sharp" turn going to Pad B. For those unfamiliar with the crawlerways at Kennedy Space Center, the two rock surfaced roads split about four miles from the

Vehicle Assembly Building. The crawlerway to Pad A continues toward the ocean, and the leg to Pad B heads northeast for a mile or so. This "brilliant scientist" was so convinced he had the answer and voiced his opinions to many of the media and others capable of influencing NASA. NASA felt forced to finally give in to his request for review and data. We allowed the man to come to Kennedy, inspect our hardware and facilities, review a lot of data, and even be briefed by KSC engineers who certainly had a lot more critical work to do. Of course, he went away with new credentials on his resume which he could use to sell his consulting services elsewhere!

Another brilliant scientist, a former employee of a NASA contractor, volunteered to tell NASA precisely what part on one of the Shuttle's three main engines had failed and caused the accident. He wanted to convince NASA and the investigative boards that a hydrogen fire in the Orbiter's main engine compartment was the cause of the explosion. I remembered this gentleman well from my engineering days on the Apollo program. He was the same alarmist who constantly warned NASA of how the Lunar Module's guidance computer would not do the job. He strongly predicted the Module would crash-land on the lunar surface, killing the two man Apollo crew.

Another promising commentator got a lot of press from telling anyone who would listen that NASA flight directors caused the accident by throttling down the main engines, then throttling them back up to maximum thrust. Of course, main engine throttling is a normal, software driven phase of all Shuttle ascents. Challenger's engine throttling was right "by the book." But this man had a scenario, he heard some air-to-ground voice, and he had the dilemma solved.

Probably the most bizarre solution to why Challenger occurred really got my dander up! Some woman provided asinine rationale to try to connect mission mishaps with crew diversity. She implored NASA to cease from flying women entirely and to carefully consider the practice of including minorities on future Shuttle crews. What a treatise against our society to even permit someone with repugnant ideas to be heard in print in our newspapers! I only mention her racist remarks here to demonstrate the absurdity of some of the correspondence we received following Challenger.

I personally received over a dozen letters proposing either why Challenger failed or a complex plan to redesign the Shuttle so it would never experience failure again. Some of the correspondence contained elaborate drawings and procedures from people who felt they had the answers NASA needed. They were simply addressed to the Launch Director at Kennedy Space Center. Most of them used my name. I suppose, to a lot of people, the Launch Director appears to the outside world to be the one person who is responsible for everything associated with a launch. Of course, the one or two redesign ideas I considered to have some valuc were passed along to the appropriate design center.

Many critics were cruel in their accusations regarding the cause of the Challenger accident. One highly decorated and recognized gentleman did well on the lecture circuit speaking on the "criminal cover-up in the Challenger disaster." Another high school science teacher claimed to have been a finalist in the Teacher in Space selection process, to have known Christa McAuliffe well, and gave many compensated speeches about his closeness to the Challenger disaster. Most of his claims were fabricated to promote his own self-image and to solicit speaking engagements across the country. Such misrepresentations were the most upsetting to accept in the months after the accident.

Several writers felt that NASA was overeager to launch Challenger on January 28 so that President Reagan could mention the Shuttle program in his State-of-the-Union address. He was to mention that an American school teacher was circling the earth as the first civilian in space. In truth, we had no pressure to launch whatsoever! On the earlier scrub deliberations on January 25, I was told that Vice President Bush might possibly stop at the KSC on Sunday, January 26 on his way to South America. At the review on January 25, we decided not to try for a Sunday launch attempt due to the weather forecast for that day. In our deliberations, the Vice President's possible visit was not even a minor consideration. Neither was any consideration given at my level to whether the President wanted to mention anything about the Shuttle in his annual speech to Congress.

There were even internal rumblings within the NASA family. Astronaut John Young, never one to be shy about speaking out, issued one of his famous "John Young memos." Although John always addressed them only to his immediate bosses, they immediately became universally distributed across NASA. The letter was highly critical of NASA management of the Shuttle program, discussed how so much of the program's safety had been compromised, and alleged that the astronaut corps did not have a voice in the critical decisions being made by Shuttle management.

My respect for John has never diminished; he is, in my opinion, one of the most renowned of America's aviators. But his timing has often brought more anguish and damage than intended improvement. Unfortunately, several astronauts echoed John's concerns, even Sally Ride, and many stated they did not want to fly again until the Shuttle program was revamped. What was so obvious to the working level troops, those who felt the anguish most, was that in no way would NASA ever attempt to fly the Shuttle again until we were totally safe. I am a strong advocate of hearing knowledgeable, experienced managers speak out when they have concerns. I would not have John Young be any other way! But there have been many instances, such as shortly after Challenger in early March, when a lot of us wished John had consulted an advisor, public relations person, or an editor in releasing his letter which one high level official described as a "poison pen memo."

John's comments were certainly valid and coming from a person of John's stature, heaped coals upon the fires already burning to correct the gross shortcomings of the Shuttle design and management control system. John's memo said "There is only one driving reason that such a potentially dangerous system would ever be allowed to fly, launch schedule pressures." John, of course, was referring to the O-rings on the Solid Rocket Motors. I have read numerous case studies and several books on why the Challenger accident occurred. There are so many who blame the management structure and schedule. In the very basic sense, a severe design fault and a poor judgment call by a few men in highly responsible positions led to one of our nation's darkest events. Still it is certainly arguable that had there been no strong schedule pressure, the Marshall management chain would have felt more inclined to push the problem “upstairs” to a higher decision level.

The Challenger seven were remembered and commemorated around the world. Among the honors bestowed upon the fallen crew members included naming seven asteroids far from earth after them. In mid-March 1986, the International Astronomical Union named asteroids numbered 3350 through 3356 after each of the Challenger seven. In our own county of Brevard near Kennedy Space Center, the school board voted to rename three local schools for members of the crew. A middle school was named for Ronald McNair; an elementary school was renamed Challenger Seven Elementary; and an elementary school was named Christa McAuliffe Elementary. Across America, Challenger and its crew were honored in similar fashion, probably more so than any other event in history.

One of the most depressing duties of the post-Challenger days involved the appearance in Washington before the Rogers Commission in the Dean Acheson Auditorium at the State Department in Foggy Bottom. Dick Smith, Arnie Aldrich, and I were called to testify on February 27, 1986 regarding the launch decisions and our foreknowledge of the “O” ring problems. On February 26, we flew to Washington, again aboard our own NASA 4 aircraft, barely arriving ahead of a late winter snowstorm that eventually closed National Airport. We were somewhat apprehensive, but by no means afraid of any questions we would be asked. I must say that I never heard anyone express an opinion that they were treated unfairly or were harassed by the Rogers Commission panel members. They treated us as would be expected from so notable a group of scientists and experts in various fields.

The trip to Washington really turned sour for me when we entered the crowded State Department building amid a wave of television and newspaper personnel. The Rogers Commission hearings were covered from every angle as the Challenger was a newsworthy subject for Americans. I was so pleased to see good friend Harry Kolcum’s smiling face at the front of the sea of reporters who were separated behind a roped area at the rear of the auditorium where we were to be interviewed by the panel. Harry was a space icon, a man loved by everyone who knew him. If there was ever a true space fanatic, Harry Kolcum, an *Aviation Week and Space Technology* reporter, epitomized that moniker. He was our pal, our supporter, but most of all, our friend.

But my attitude quickly changed negatively when I saw Dr. Rocco Petrone and three of his top Rockwell managers huddled together. We had been alerted that they would testify that their company opposed the Challenger launch and never gave a "go" call to commit to proceeding that fateful day. I could tell by the way that two of the managers shunned eye-to-eye contact that they were not there to praise NASA. Later, Petrone told the panel that he passed along to Robert Glayser, his representative at the launch site, that Rockwell could not recommend proceeding with Challenger's launch. Bob Glayser told the panel that at a 9 a.m. ICE team meeting, he raised the question to Arnie Aldrich that he could not be sure about how the ice on the pad would impact the Orbiter's tiles. He testified that his statement at the meeting a couple of hours before Challenger's launch was basically, "Rockwell cannot assure it is safe to fly." Later on that day, Arnie told the panel that Rockwell surely raised a concern, but never did they recommend that we not launch Challenger.

Never had I been so disappointed in a company, an individual of Petrone's stature, or a group of individuals like the Rockwell party of four. I considered this appearance before the Rogers Commission to be nothing more than an attempt to protect their company from liability in being part of the Challenger accident. This was a pitiful display of "stretching the facts" to avoid financial losses and political embarrassment for their company should litigation be forthcoming. I have been told by insiders at Rockwell that Petrone was only venting his frustration at not having been able to stop the Challenger launch countdown, that he was being himself and seeking some degree of vindication! I did not read that in the voices, the facial expressions, or the behavior of any of the four witnesses. I shall forever consider this to be a "low blow" against NASA, and the programs in which Rockwell was a major participant. Of course, my interpretation of Dr. Petrone's testimony is my own personal opinion, not an official NASA position. I salute the three or four other major aerospace contractors who stood proudly and took the criticisms that came their way. All of them were especially conscientious to find their faults and correct them. The Thiokol Company, in particular, should be praised for its responsiveness to the intense effort following Challenger to correct the O-ring seal problems. Thiokol worked hand in hand with NASA to facilitate a much safer, more reliable Solid Rocket Booster joint redesign.

I also fully remember the final poll which was made of each of the major Shuttle contractors late in Challenger's countdown. On the day of Challenger's launch, Rockwell was polled and clearly gave an unqualified "go" for launch. I heard that strong affirmation of Rockwell's "go" status, and it was recorded as essential voice data for the countdown. No contractor has contributed more to America's major space programs over the last thirty years than Rockwell, formerly North American Aviation. Thousands of loyal Rockwell employees have given their lives and careers to the space program. They are my friends and comrades. For no other reason than respect for Rockwell's outstanding workforce, I wish that Mr. Petrone and his managers had not opted to appear before the Commission with their testimony. I have a comfortable feeling and peace of mind that the astute minds of the Commission members were quite aware of the intentions of Rockwell's leaders.

During our testimony before the Commission, Chairman Rogers asked Dick Smith, Jesse Moore, Arnie Aldrich, and me if we had any knowledge regarding Thiokol's objections to launching the Challenger. All four of us replied either, "I did not" or "No, sir".

Mr. Hotz further questioned me about the launch temperature.

Mr. Hotz: "Mr. Thomas, you are familiar with the testimony that this Commission has taken in the last several days on the relationship of temperature to the seals in the Solid Rocket Booster?"

Mr. Thomas: "Yes, sir, I have been here all week."

Mr. Hotz: "Is this the type of information that you feel that you should have as a Launch Director to make a launch decision?"

Mr. Thomas: "If you refer to the fact that the temperature according to the Launch Commit Criteria should have been 53 degrees, as has been testified, rather than 31, yes, I expect that to be in the LCC. This is a controlling document that we use in most cases to make a decision for launch."

Mr. Hotz: "But you were not really very happy about not having had this information before the launch?"

Mr. Thomas: "No, sir, I can assure you that if we had had that information, we wouldn't have launched if it hadn't been 53 degrees."

Obviously, the stress of the recovery phase following Challenger provided us with a lot of ups and downs. For many days directly following the accident, there were a lot of sad faces and dejected employees at the KSC. Then it was splendid to see our Kennedy launch team and the thousands of support personnel roll up their sleeves and attack the actions we were assigned. There was better cooperation and teamwork among the three major Shuttle centers (Johnson, Kennedy, and Marshall) than ever before. Astronauts were assigned to work with engineers and managers on a daily routine basis. We at Kennedy were pleased to see Hoot Gibson, Steve Hawley, Charlie Bolden, Bo Bobko, Brewster Shaw, Bob Crippen, and other astronauts as part of our role in the investigations. We tried to assure that the Shuttle launch team morale and proficiency remained at the highest possible level. On October 9, 1986 we rolled a Shuttle stack with the Orbiter Atlantis to Pad B. Led by STS-61C Commander Hoot Gibson and Pilot Charlie Bolden, we practiced a simulated countdown which Hoot later described as a "real morale booster." It was extremely encouraging to hear and feel the performance of the launch team at its very best. I remarked to the press, "Whenever we have a vehicle that we and the program feel is safe to fly, KSC will be ready to launch. The launch team is just as sharp as ever."

We later ran a full-scale pad emergency rescue simulation with the help of an astronaut crew led by Frank Culbertson and Steve Oswald. Simulating an "observed fuel spill," we "tagged" four of the Orbiter closeout team as incapacitated, including Astronaut Culbertson. We sent a fire rescue team to the Orbiter level of the pad structure to rush the "victims" to an evacuation helicopter. Again, all aspects of the KSC operations were outstanding. A well-rehearsed, enthusiastic launch team proved itself ready for return to flight.

A few days before Thanksgiving we rolled Atlantis back from the pad. I was glad to see morale improve as we all felt that the approaching first anniversary of Challenger would be hard to cope with. We observed the first anniversary of what has since become traditional, a 60-second pause for silent prayer and remembrance at 11:38 a.m. I found a way to sneak back to my seat in the firing room and sat there in silent prayer to God as I recalled that sad day. I don't know why I wanted to be there on that one occasion, the first anniversary. I don't think I ever again sat in the Shuttle Launch Director's chair. I knew I could never serve as Launch Director again!

In December 1986, shortly after Atlantis' successful pad testing, the Kennedy Space Center announced a major restructuring of its management team. Bob Sieck was again returned to the job of Shuttle Launch Director, a position which demanded someone of Bob's caliber. Tom Utsman assumed the top Shuttle management position as director of all Shuttle operations and engineering functions. The Kennedy Space Center had a new leader, a gung-ho three star General Forrest McCartney. General McCartney brought new order and direction to a proud center. A crew-cut career officer, he commanded respect and got it. We rallied around our new leader, and I don't believe anyone could have done a better job than McCartney in getting the KSC team back together. He became a friend to all Kennedy employees and encouraged them to love their work. General McCartney always promoted the idea that we should "whistle when we come to work and whistle when we go home."

I was pleased when the general asked me to reorganize the center's safety, quality, and reliability operations into a first line directorate equally sharing launch readiness responsibilities with the engineering and operations organizations. Safety and quality had been fragmented across a lot of smaller organizations, and KSC seriously needed a "check and balance" function in these two crucial disciplines. It was no doubt the greatest managerial challenge I had ever encountered. But with a strong recruiting effort, I enlisted some of the best Kennedy people to assume key slots in the new organization. Our new safety and quality organization grew to become the standard bearer for all of NASA.

In September of 1986, NASA announced the formation of a Space Flight Safety Panel which would have oversight responsibility for all NASA manned space program activities affecting flight safety. The first chairman selected with Astronaut Bryan O'Connor who had been the pilot on STS-61B, which I had served as Launch Director for in November of 1985.

NASA re-engineered itself. NASA purged itself of some of the practices that crept into its ranks and caused it to become "bureaucratic." After a total reconstruction of our test processes, requirements, and procedures and a total redesign of the solid motor joints, we again launched the "new" Shuttle into orbit. On September 29, 1988 a crew of five commanded by Rick Hauck flew Discovery on what we named STS-26R, the R designating "redesign" or "return to flight." Flying in the pilot's seat was Dick Covey who had been the capsule communicator on the day Challenger exploded. The new solid motor joints performed flawlessly. The Shuttle was back out front again as the world's only manned access to space. We were progressing slowly with America's international space station which depends primarily on the Shuttle for its construction, re-supply, and manning. I confidently believe the Shuttle will continue to be America's only manned access to space for at least the first two decades of the 21st century.

Although disasters are horrible events to torture the human spirit, often *from the ashes of disaster grow the roses of success.*

CHAPTER 25

The Rogers Commission

What we got here is a failure to communicate.

From the movie *Cool Hand Luke*

The final decision by the Reagan Administration to form a blue ribbon investigative panel to investigate the Challenger accident was not made until January 31, 1986. White House Chief of Staff at the time, Don Regan, finally persuaded President Ronald Reagan that NASA should not be allowed to investigate itself. A lot of NASA proponents, former NASA Administrator James Fletcher, Congressman Bill Nelson of Florida, and Senator Jake Garn of Utah were strongly vocal against an outside group being considered for investigating NASA. The final decision to charter an outside group was made by the President on the presidential plane returning from the Challenger memorial service in Houston on January 31.

Executive Order 12546 dated February 3, 1986 created the Presidential Commission on the Space Shuttle Challenger Accident. It was to be generally referred to as the Rogers Commission. The commission was given 120 days to review the accident and report to the President by June 6, 1986. The commission stuck to its deadlines and actually presented its report to President Reagan on June 9, 1986. This was a tremendous accomplishment, considering the magnitude of the data and complexity of the Shuttle program.

President Reagan in announcing the commission said "The commission will review the circumstances surrounding the accident, determine the probable cause or causes, recommend corrective action, and report back to me within 120 days."

There are certainly no innovative remarks in these directions. These are standard instructions which have been given to hundreds of investigative groups that I have either served with or supported.

The commission members received no pay for serving; only their direct expenses were compensated.

Regan, the busy Chief of Staff, also found time to personally select William Rogers to chair the panel which was to become the Rogers Commission. A loyal Republican, Rogers served as Attorney General in President Eisenhower's Cabinet and as Secretary of State under Richard Nixon. A lawyer by profession, Rogers was known to be fair but a real "details" man, one who strives to be factual. Rogers was at the time a partner in the New York law firm of Rogers and Wells. A no-nonsense Chairperson, Rogers was reported to be annoyed with Nobel prize-winning physicist Dr. Richard Feynman, calling him an "iconoclast." Feynman, who was a participant in the early atomic bomb experiments, provoked a lot of intelligent NASA scientists by doing a "school boy experiment" before the commission and the media. I think Dr. Feynman, a brilliant theorist, was far out of his element when dealing with the NASA scientists and engineers who were accustomed to transforming theories into real operating hardware. Feynman placed a piece of rubber in a glass of ice water to show how cold temperature will change the elastic properties of a silicone product such as the solid rocket motor's O-ring seals. This amazing doctoral revelation was evidently considered by Feynman to be the Sherlock Holmes discovery that solved the dilemma. Actually, NASA had already briefed the commission on the O-ring problems. It was by that time quite common knowledge that the O-rings had failed to protect the joint from emitting the horrendously hot gases.

Among the other name members of the panel were Sally Ride, America's first woman in space, and Neil Armstrong, the first man to set foot on the lunar surface. President Reagan chose Armstrong to serve as Vice Chairman of the commission. A notable member but one who was to take little part in the investigation was famous test pilot, Chuck Yeager. Yeager was more conspicuous by his absence from the commission deliberations rather than his contributions.

The Rogers Commission, in spite of its famous membership, did a very commendable job. Although the members would give the impression that they were chosen because they were renowned rather than for the contributions they could make to such an important investigation. They organized into subgroups according to the issues they wished to address. Bob Crippen, Brewster Shaw, John Fabian, and other astronauts were asked to take important staff positions. Crippen signed out an agenda for a Rogers Commission subpanel visit to KSC on March 4-5, 1986. The agenda covered launch team training, documentation, workloads, schedules, processing, recovery status, and included a review of launch photos and television coverage. It was a hectic two days, as I was closely involved as the lead KSC manager for several of the review sessions. The panel asked probing questions. I was extremely proud of how the KSC team was prepared. We answered the inquiries well and earned a lot of respect from the panel.

Approximately two weeks into the Rogers Commission's deliberations, they announced that the decision to launch Challenger "may have been flawed." They directed that individuals involved in the launch decision no longer take part in NASA's internal investigation. President Reagan was reportedly told of the commission's decision. The decision, of course, included those who had worked through the Challenger countdown to clear the numerous problems we had faced. Jesse Moore, Arnie Aldrich, Center Directors Dick Smith and Bill Lucas, Bob Sieck and I were all included in the launch decision process. I never once felt a let-up in my post-Challenger work activities. They remained time demanding and required a tremendous amount of energy.

The March 3, 1986 issue of *TIME* dramatically reported the results of one of our meetings held at the KSC with a Rogers Commission panel as follows:

"There was a stunned silence in the commission's closed hearing room at the Cape after Bob Sieck, Shuttle Manager at the Kennedy Space Center; Gene Thomas, the Launch Director for Challenger at Kennedy; and Arnold Aldrich, Manager of Space Transportation Systems at the Johnson Space Center in Houston all testified that they had never before heard that Thiokol engineers had objected to the launch. Rogers ordered everyone except the commission out of the room and declared "We must advise the President as soon as possible." Explained one commission source, "We did not want the President to be blindsided."

I honestly don't remember this meeting ever taking place. I was either not present or not privy to Chairman Rogers' direction.

The same March 3 issue of *TIME* reported that Jesse Moore testified before a Senate subcommittee that he had not been told of Thiokol's opposition to the launch or discovery of cold spots on the solid rocket boosters.

A young Tennessee Senator, Al Gore, Jr., was quoted "The record calls into question the way alarm bells are rung and heard" at NASA.

This same Senator Gore on February 22, 1986 had called for NASA Administrator Dr. James Beggs, who was on leave, and acting Administrator William Graham to resign. Gore felt that Graham had misled the Senate Space Committee during a hearing about how NASA's top managers knew about the solid motor O ring seal problems. "Safety was given a back seat to the requirements of the launch schedule," Gore said. "They were so eager to get the Shuttle off that day that they ignored clear warning bells."

There were a lot of "discoveries" as we plowed through the thousands of documents used for each flow of a Shuttle mission through the KSC. I was not pleased to see some of the results, although they were not contributors to the cause of the Challenger accident. As a manager, you must rely upon a large number of other people to make a cumbersome process workable. Those of us who worked for the government could only oversee our contractor workforce at the top levels. We were too few to probe deep into the details of the day-to-day operations. In this atmosphere, it is easy to be misled or to assume an operation is being performed in a satisfactory manner. Two particular findings which I found upsetting were uncovered as we reviewed our processing of the Shuttle for the Rogers Commission.

The first finding really raised a red flag. We were tracking contractor workers' overtime data and were pleased to see the overtime per week around 20%. In our estimation, 20% overtime amounted to 8 hours per week or a 6 day work week. In reality as we plunged deeply into how the contractors accounted their overtime, we found that they simply divided the total number of hours of overtime accrued per week by the total number of employees in a work unit. So even employees who never worked any overtime were counted. So were clerks and secretaries. The resulting figures of 20% or so overtime were actually hiding a gross misuse of overtime by critical Shuttle workers. Some engineers and technicians were working 60 hour weeks as a standard practice. The amount of overtime worked often exceeded 20 hours per week. I was especially incensed to have had the overtime calculated in such an outrageous fashion. I can only blame myself and our NASA workers for being blind to this situation. We were concentrating on schedules and demands of the hardware processing more than paying attention to details. We learned an important lesson from this data. We adapted a version of the overtime policy used by the Nuclear Regulatory Commission, a high-tech hazardous agency, whose work is similar in nature to that of NASA. Today, overtime usage is monitored on a regular weekly basis. High level management approval is required before a Shuttle worker is allowed to exceed a clearly defined and well documented set of overtime guidelines. This disciplined overtime policy can be partially attributable to the reviews conducted by the Human Factors subpanel of the Rogers Commission.

The second disappointment from the findings concerned the laxity in quality control of the documentation we used to process the Shuttle flight hardware. We reviewed every one of the hundreds of test and checkout procedures. We reviewed millions of lines of instructions to the workers in the Operations and Maintenance Instructions. These OMIs were our bread-and-butter, our cookbooks, our operations manuals. Fifty percent of the line item instructions required either a contractor or a government inspector, or both, to witness the work as it was done. Imagine a housewife cooking a great loaf of banana bread by an old family recipe written on a 3 by 5 index card aged by butter and oils. Imagine this same housewife having the next door neighbor housewife standing by her side in the kitchen watching her measure the ingredients, set the temperature on the oven, mix and stir the batter, and note the time the sweet-smelling concoction was placed in the oven. As each step is performed, the lady from next door initials the line-by-line instructions on a copy of the recipe. If the husband's boss is expected for dinner and the banana bread needs to be especially yummy, the cook may be required to have two housewives watch her prepare and bake it. A silly analogy you might say! But that's how quality coverage is accomplished in most high-risk operations where the slightest error may cause disaster. We considered our quality to be the best there was. And I would still argue its merit and defend it against any other similar independent assessment program.

But we found a lot of instances where quality control verification stamps, both those of our contractors and our NASA quality inspectors were missing. The work had been done, the technician stamped the work instruction signifying his completion of the procedural work steps. But in too many cases the quality control "buys" were either blurred beyond identification or non-existent. In a probability-based environment, one has to accept a certain degree of error, a somewhat predictable amount of mistakes. We were not willing to accept any oversight and negligence with the critical procedural instructions associated with the flight elements. Someone has said "to err is human." We did not expect super-human performance and knew we could never attain a zero-error atmosphere. But we wanted to protect the safety of the Shuttle crews and the flight hardware without compromise. After I saw what I considered to be far too many missed quality verifications, I was one of the many who insisted on adding additional mandatory inspection points to the new procedures we wrote in the Shuttle return-to-flight timeframe. It is certainly ironic that we were strongly criticized for our burdensome paperwork systems by the media, the Rogers Commission, and by groups such as the Congressional Investigation team. And the only way to respond to some of their recommendations was to further encumber the paperwork systems.

During a February 27, 1986 public hearing on the Challenger accident, Chairman Rogers said: "The trouble with so much paperwork is you eliminate the element of good judgment and common sense."

I totally disagree with his choice of verbs. Had he chosen to use "diminish" or "reduce" instead of "eliminate," I would strongly agree. There is probably nowhere in the world where a thinner line exists between the use of too much paperwork and not using enough paperwork than in the space program. I'm certain we overreacted after Challenger and increased our documentation way beyond what was necessary. We also re-assessed the test requirements for certifying the Shuttle flight hardware elements. This list grew tremendously as design engineers felt a pinch to totally certify their systems prior to committing to fly again. The Rogers Commission hearings put a lot of pressure on the engineers to reestablish their ideas of rigid testing with little regard for schedules or costs. The return-to-flight phase of the Shuttle recovery complemented this test philosophy. But as the Shuttle began to fly safely on a regular basis, the overreaction which added more testing, more thorough documentation, and more mandatory inspections was to become a nightmare for the Shuttle budgets! Shuttle managers must cut costs! Cut the cost drivers! And the engineer who was encouraged to be sure he tests his hardware properly before flight can't believe he has been asked to cut back. This pattern of overresponsive reaction to the findings and recommendations of an investigative panel is standard operating procedure in NASA.

The Rogers Commission was totally justified. NASA should not investigate its own shortcomings! The Commission ruled that the launch decision process for the Shuttle program was "flawed." The use of the adverb "flawed" seems to have originated with Chairman Rogers himself. I would take exception to the process being "flawed." Had everyone followed the rules, the process certainly allowed for reporting of the O-ring controversy between the solid rocket motor builder, Thiokol, and the government customer, the Marshall Space Flight Center.

The commission served as a forcing function to cause the right people to make the necessary changes, both in hardware design and in processes to correct the Solid Rocket Motor joint O-ring failure mode. The Commission made little discovery on its own. Dr. Feynman's brilliant physics lesson was not a revolutionary breakthrough. With the Commission's big and laborious report came long hours of rework and recertification of every significant piece of flight and ground hardware. I feel that it was a necessary process that we needed to work through in order to gain the confidence in ourselves to again commit astronauts' lives back to Shuttle flights. For that very purpose alone, the Rogers Commission Report was worthwhile.

The newly appointed NASA Administrator Dr. James C. Fletcher stated, "The report of a presidentially-appointed, independent body carries with it special status and the compelling obligation to study its conclusions with great care. We are prepared to do that with an open mind and without reservations."

He also strongly promised to return NASA to be the proud agency that it had historically been noted for when he said, "We are going to behave like a family which has suffered a tragic event. We are going to deal responsibly with our loss, without needless recrimination, and we are going to move forward, facing and conquering the challenges that face us. Where management is weak, we will strengthen it. Where engineering or design or process needs improving, we will improve them. Where our internal communications are poor, we will see that they get better. This is an agency whose excellence and commitment to new frontiers drew to it seven exceptional Americans...."

The Rogers Commission served a significant role in bringing the Shuttle program back into reality. The members of the commission that I worked with toiled long and tediously to get a handle on NASA's technical and managerial problems. Their technical expertise added little. The Marshall and Johnson Space Centers design engineers and their contractors did an outstanding job in redesigning the solid rocket motor joints. Three levels of redundancy were implemented in the O-ring redesign. All Critical Items were revisited for each of the flight hardware elements. An escape system similar to a fireman's pole was added to the Orbiter. Heaters were added to the motor joints to allow ground controllers to keep the O-ring areas warm in freezing temperatures. Numerous hardware changes were made that increased the factor of safety and reduced the risk of another catastrophe. Little of the technical design enhancements were driven by the commission's findings.

In contrast, the commission made a lot of recommendations regarding NASA's communications, organization, chain-of-command, launch decision process, and safety practices. We expected no less from a strong panel of former aerospace managers and pilots. Chairman Rogers did an exceptional job of channeling the commission's focus and keeping them on schedule. America owes the commission a debt of gratitude for conducting a disciplined and in-depth review, yet not destroying the pride of a great agency. The commission's recommendations brought about numerous managerial and process changes. Some in NASA felt they were excessive. My own career took a turn in the direction of management based on the commission's recommendation to increase the safety and quality oversight. I was asked to become the KSC Director of Safety, Reliability, and Quality Assurance. I took the new job gladly. The commission's direction to empower a strong independent safety and quality organization was taken seriously by NASA. Soon the safety and quality function was given equal authority and responsibilities as had always been held by engineering and operations.

Many changes were brought about due to the credibility and esteem associated with a presidentially-appointed investigative commission. I consider them all to be appropriate and productive. However, two results of NASA's restructuring have negative connotations. In many areas, NASA overreacted, causing a return to stability and reasonableness to be difficult. Secondly, the attempts to bring about changes and moderate these overreactions left NASA open for severe criticism. I have heard a lot of bad-mouthing from both inside and outside the agency that NASA reneged on its commitment to carrying out the commission's recommendations. This almost subjects an agency to be in a "lose-lose" situation. Overreaction to investigative findings burdens a program with unnecessary requirements. Relaxation of new stricter control functions opens the door for criticism of your commitment to keep safety as the number one program priority. Such is the environment that evolved for NASA after we implemented the recommendations of the Rogers Commission.

What we got here is a failure to communicate. I was at first somewhat reluctant to use the quote from the movie classic, *Cool Hand*

Luke. I was concerned that I might be too cavalier in dealing with the cause of the Challenger accident, truly one of the most disheartening events in America's short history. But as I pondered more and more over this short pronouncement used by the prison warden of Lucas' chain gang, I felt that it succinctly sums up the primary failure causing the Challenger disaster. We had failed to communicate, simply and assuredly! The underlying tragedy in this drama was that a sinister malfunction, the O-ring characteristics, was the message we failed to communicate! It was a message that could have saved seven great people had it been communicated properly!

CHAPTER 26

How Does a Christian Cope?

> *For I am persuaded that neither death nor life, nor angels nor principalities nor powers, nor things present nor things to come, nor height nor depth, nor any other created thing, shall be able to separate us from the love of God which is in Christ Jesus our Lord.*
>
> Romans 8:38-39

I almost entitled this last chapter "The Most Important Chapter in This Book." Some readers will undoubtedly comment, "He captured the insider's essence of the Challenger accident, and then he went off on a religious thing."

I dearly trust that throughout my accounting of the Challenger story I have demonstrated that it is through abiding faith and amazing grace that God has blessed my life and the lives of my loved ones. I also trust that I can express through words the ideas of my heart. I pray that everyone who reads this book would read these closing words with a mind towards what I consider the real purpose of our lives on earth... to seek and worship the Almighty. And when times of trouble and chaos arise in other lives, I would pray that they would find the comfort of knowing that God is still in control of this universe!

The primary purpose behind my desire to document my involvement with the Challenger accident was not to record historical events. The message I would deliver is simply:

A Christian can cope with disaster through the power of the Holy Spirit.

As the Shuttle Launch Director, I was on top of the world! I had the very best job in the entire world. It was challenging! It was rewarding! I had been successful in my career and had risen through the ranks to the best position in the space program. NASA was launching the Shuttle regularly; we would soon launch from the west coast launch site; we were designing a Space Station!

Things were great at home! My oldest daughter was engaged to a fine young ministerial student. My son was doing well in college on his way to becoming a youth minister. My youngest daughter was an honors student in high school. My sweet wife and I struggled to provide a Christian home and the best for our family. God had blessed us with all we needed: a fine family, a church environment where we were close to other Christians and where Christ was exalted, and a job too awesome to describe. I was truly on top of the world. What an exciting, exhilarating career! I thought I wanted to be Launch Director forever. I didn't want to go any further in management. I had found my niche, and I was extremely comfortable in this super job. I was leader of the greatest launch team in the universe.

Suddenly the serenity of the world's best job was shattered 73 seconds after the Challenger spacecraft lifted off from Kennedy Space Center on January 28, 1986. My whole world and everything I cherished seemed to be totally obliterated in a moment. A white cloud of an explosion, reddened in places by hypergolic fuels and two wayward rocket boosters streamed away erratically, almost forming the sign of the cross. With that gigantic fireball, I lost dear friends; I lost some degree of pride; I lost considerable confidence; and I almost lost my faith.

Looking back in retrospect to why Challenger happened when I was Launch Director, I recognize that Jesus said, "He makes His sun rise on the evil and on the good, and sends rain on the just and on the unjust." (Matt. 5:45) By no means do I purport to be good or just, but I also felt I wasn't involved in evil or being unjust. God has

His plan for dealing with history, and I want to someday understand why things good and things bad happen as they do.

I know assuredly that I could feel the voice of God speak to me through the Holy Spirit which comforted me in the first few minutes after Challenger exploded. This same Comforter was felt by close members of the crew's families, and through other Christians like my son Chuck. I know God assured me that He was still master of the universe and still in control of the situation. This is much the same message God spoke to Chuck in chapel when he first heard of the tragic accident.

But as the minutes became days and the shock of the event began to subside, the reality of the impact of such a disaster became so definite. Amid the tears and woeful cries of "Why, Lord, why?" I began to ponder deep in my innermost heart why I had been so close to a fateful event in history. Why was someone who never wanted to be close to the headlines now one of the hundred or so people who had been a party to a launch which would be studied and second-guessed as long as the space program continued? Was it God's plan that I experience this trauma? Since my childhood acceptance of faith in Jesus, I had depended on God's grace for every walk of my life. Had this been God's grace still at work in my life? I sincerely admit I did not know. I was somewhat confused and had little time for deep thoughts as we were so extremely busy performing the investigations and getting the recovery operations ready to fly again. In reality, this busy period was a blessing in disguise.

When I did go about a process of thinking through the doubts and questions of why Challenger had to happen, I asked myself in so many words a question strongly delineated by the scripture of Romans 8:35: "Who shall separate us from the love of Christ? Shall tribulation, or distress, or persecution, or famine, or nakedness, or peril, or sword?"

I quickly realized that just a few lines in the Bible were to sum up specifically how I managed to cope with the Challenger accident. I was inspired to read again and again the blessed assurance of Romans 8:38 and 8:39: *For I am persuaded that neither death nor life, nor angels or principalities nor powers, nor things present nor things to come, nor height nor depth, nor any other created thing, shall be able to separate us from the love of God which is in Christ Jesus our Lord.*

I knew beyond any possible doubt that my faith in Jesus Christ was real. I knew that the love of God was dear to me and I had always lived by faith.

Now afresh and anew, these two verses of divine Scripture assured me that there was absolutely nothing that could separate us from the love of God. What a tremendous source of strength and encouragement! Nothing living or dead, nothing above or below, nothing real or imagined, no thing, angelic or earthly, could ever separate me from God's love. If every professing Christian only accepted that persuasion, our world would be miraculously changed. I never recall the Challenger tragedy without immediately having these two Holy Scriptures come to my mind to assure me of the inseparable grace of God.

It is quite probable that every Christian will face a day of destiny much as I experienced with the Challenger. I was emotionally shaken as many Christians will be when devastating circumstances happen in life's journey. How can we cope with these disturbing events? How does a Christian cope?

How did the men and women close to Challenger cope? Many of the top level managers were purged and lost their jobs. Some took it very hard; some even turned to alcohol to drown their woes. Some became bitter at others, the Space Agency, and at themselves. There were those who just quit - pulled up stakes and moved on! Some chose retirement which for many was long overdue. Some company executives rose to defend their names and honor. Some humbly accepted blame and responsibility. Critics abounded, some with good credentials, some with none. Some wonderful Christian soldiers marched onward and kept the faith. Through the assurance of God's word, I kept the faith to stay the course and help bring our country's space program back to its glory days.

Compared to other great memorable events, the Challenger accident was a historic tragedy of significant proportion. Along with the assassination of John Fitzgerald Kennedy and the Apollo 11 landing on the moon, Challenger ranks near the top of historic memories of the 20th Century.

But Challenger and all other tragedies pale when we think of the importance of eternity. Nothing is more significant. I would pray that the men and women who dedicate their entire lives to space would similarly dedicate their minds to understanding the reality of eternal life.

In the fall of 1995 I was asked to speak at an Episcopal church in Titusville, Florida, which adjoins the Kennedy Space Center. The church was without a pastor and needed someone to speak at a Sunday morning worship hour. I gladly accepted the invitation and chose to speak on how, as a Christian, I had found peace with the Challenger accident. The Shuttle launch scheduled for Friday slipped into that Sunday morning time frame. I realized I had to either miss only my second Shuttle launch or back out of the speaking engagement. I decided to remain faithful to the church commitment and told the other NASA managers I would miss the launch. As we were preparing for bed on Saturday evening, I told Juanita, "I cannot believe that they are counting down a Shuttle launch, and I am not out there in the Firing Room tonight." As wives often do, she spoke with wisdom and replied, "Fifty years from now, they will still be launching space missions, and it won't matter one bit to you!" What a remarkably true statement of the fact that life is but a short episode, a small tick of the clock of eternity. Fifty years from now, man will still be launching space machines to foreign planets, still be trying to keep peace, still be preparing to make war, still be trying to cure terrible diseases, still be trying to make more money and enjoy life better. All this effort is of no significance in the eternal destination of every living soul. As I mature as an individual, I more and more recognize the brevity of life and the certainty of the afterlife. Billy Graham has said that one of his greatest observations was the briefness of life, how fast our lives go by and how short our tenure on earth is. I have learned that time slowly heals the tragic shock of the Challenger experience.

Still, few days pass in which I do not recall Challenger. Sometimes the thoughts are sad, often just reminders of events of that day; sometimes there is peace about the outcome of the incident. And in all of this, I realize it won't matter when we spend eternity with a living God. Fifty, one hundred, a thousand years in eternity with Him are all that matters.

I am often reminded of God's eternal love towards us and the grace of Jesus Christ through a simple newspaper cartoon called *B.C.* Johnny Hart periodically presents the gospel through his comic strip. I proudly carry several of his strips in my personal Bible. One comic strip has been especially relevant as to what we have done in our space exploits. It is entitled *I'll Take Heaven* and is a poem depicting the gospel in no uncertain terms.

Man, man, magnificent man,
Creates forces that outshine the stars,
He can shoot himself up and tap dance on the moon,
Or hurl himself clear out to Mars!
He can unleash a force that evaporates steel,
Since he's learned how the atom behaves,
Yet he has no recourse but to bow to the force
That summons the dead from their graves.

I am persuaded that the force that summons the dead from their graves, the power of a risen Christ, is the only means of hope and salvation man has.

A space enthusiast not associated with space work, a minister, once remarked to me, "You must be immensely pleased with what you have accomplished and all the many things you have seen and done in the space program. It must make you feel extremely proud!" At the time, I think I simply acknowledged his kind words and thanked him for his comments. Today I am quite certain, if I had time to consider his compliment and my answer more thoughtfully, I would have used a Scripture from Philippians 3:7-8: *But what things were given to me, these I have counted loss for Christ. But indeed I also count all things loss for the excellence of the knowledge of Christ Jesus my Lord, for whom I have suffered the loss of all things, and count them as rubbish, that I may gain Christ.* As those of us from NASA make public appearances, meet with the media, and give promotional speeches, we are asked all manner of questions, some very thought-provoking and intelligent. Children always want to know how the astronauts go to the potty in space. Another popular query is whether as a NASA person we believe in life on other heavenly bodies in the universe. Do we believe aliens have contacted or even visited planet earth? Do we believe in unidentified flying objects? Even, does NASA have the body of an alien hidden away somewhere? A familiar question that I have been asked time and again, "Did you rocket scientists really send men to the moon? Or was it a giant hoax? Do you really send people into space as regularly as you are reported to do? My dear aunt died convinced that we did not actually send men to the moon during the Apollo program. She said we simply staged the ground operations by having the three-man crew go up one side of the pad in clear view of millions of television viewers. Then we escorted them out a back door exit, off the pad to an elaborate movie studio set where they pretended to land on and walk the moon's simulated surfaces. Accordingly, we always shot the huge expensive Saturn V launch vehicle to add as much deception to the facade as possible. I had numerous discussions with my aunt concerning the facts of the moon landings, but I am certain that I never convinced her that we really accomplished these marvelous journeys to the moon and back. Another dear relative Uncle Eddie once warned me, "If you space guys don't stop going to the moon, you're going to knock a piece of it off and ruin fishing around here for the rest of time!"

So how do you convince a skeptic that NASA really landed men on the moon; that twelve American males have actually bounced around on the lunar soil; that the American flag is still the only flag on the moon? Well, quite frankly, I don't argue or try to convince anyone any longer that we did what we claimed to do.

If I wanted to convince any living person about something, it would be to try to convince them of the certainty of eternity and that there is life after death. I would use every fiber of my body, every brain cell of my mind, and every bit of energy I could muster to convince them of something of eternal significance. I would impress upon every individual that we are promised eternal life through Jesus Christ if we recognize him as Saviour and accept his death and resurrection through faith and that through His grace we are saved through faith. Modern thinkers like to believe they are open-minded and super intelligent. I would contend that real wise people recognize that a greater force, a mighty God, built, reigns, and controls the vast universe.

Want to experience the supernatural? Want to speak and learn of life beyond the limited boundaries of the physical which we experience on this tiny earth? Want to continue to live as a spirit and soul after the body serves us no more? These experiences can only be known through an abiding faith in Jesus Christ.

My argument before men would not be to convince them of the reality of moon landings. I would spend all my resources in convincing them of one powerful truth. This powerful truth is so simply stated in a popular chorus that I have heard sung literally thousands of times in churches:

> *He is Lord, He is Lord,*
> *He is risen from the dead and He is Lord,*
> *Every knee shall bow, every tongue confess,,*
> *That Jesus Christ is Lord.*

Do I personally believe that our earth has been visited by other heavenly sources? Beyond a doubt.

I believe that an omnipotent, omnipresent, and omniscient God visited earth in the human body of Jesus Christ of Nazareth! To men of his time, he was truly an alien. They did not accept his message and crucified him. But being God Almighty in the flesh, he was resurrected from the grave and sits at the right hand of God. He is Saviour to all who will profess faith in Him. I believe this with all my soul. Every engineer, technician, astronaut, man, woman, government official, athlete, king, queen, peasant, every living person must someday address this same Christ. The Bible says that *At the name of Jesus every knee shall bow and every tongue confess that Jesus Christ is Lord* (Phil. 2:9-10). The technological marvels of the space program that I have written about have allowed the Gospel to be spread throughout all the earth via communications satellites. Everyone must acknowledge faith or disbelief in God's only Son. There is no middle ground, no neutrality; you either believe on Him, or you do not!

I'm convinced that if every intelligent human being would seek to understand and know the perfect will of God, they would see Jesus Christ as their only redemption, their assurance of eternal life. In times of work crises, several United States presidents have turned to Billy Graham for inspiration in seeking God's wisdom and blessing. When royalty and the famous are laid to rest, they too, like the poor and common people, look to Christianity for solace and peace. When each of us reaches the point of carefully considering what comes after this short, short period of life on earth, we must prayerfully seek the answers through the Holy Bible. We must swallow the pride within our hearts and humble ourselves before the almighty God.

How can I end an accounting of an event of such historic proportion with a plea for spiritual awakening?

With one message: *For I am not ashamed of the gospel of Christ, for it is the power of God to salvation for everyone who believes, for the Jew first and also for the Greek* (Rom. 1:16).

I look forward to seeing members of the Challenger crew in the hereafter, because our common faith in Christ has promised that we will spend eternity praising almighty God.

IN MEMORIAM

We cherish your memory, Dick Scobee.
We cherish your memory, Mike Smith.
We cherish your memory, Ron McNair.
We cherish your memory, Judy Resnick.
We cherish your memory, Ellison Onizuka.
We cherish your memory, Greg Jarvis.
We cherish your memory, Christa McAuliffe.

Now I saw a new heaven and a new earth,
for the first heaven and the first earth
had passed away. Also there was no more sea.

Revelation 21:1

Printed in the United States
133846LV00003B/226-300/A

9 781600 340963